The Italian Genius on Display

Nuncius Series

Studies and Sources in the Material and Visual History of Science

VOLUME 11

The titles published in this series are listed at *brill.com/nuns*

The Italian Genius on Display

The First National Exhibition of History of Science (Florence, 1929) and the Preservation of Scientific Heritage in Fascist Italy

By

Francesco Barreca

BRILL

LEIDEN | BOSTON

Cover illustration: Flyer announcing the Exhibition. © *Museo Galileo Archives.*

The Library of Congress Cataloging-in-Publication Data is available online at https://catalog.loc.gov
LC record available at https://lccn.loc.gov/2022011342

Typeface for the Latin, Greek, and Cyrillic scripts: "Brill". See and download: brill.com/brill-typeface.

ISSN 2405-5077
ISBN 978-90-04-29740-1 (hardback)
ISBN 978-90-04-29741-8 (e-book)

This book is printed on acid-free paper and produced in a sustainable manner.

In memory of Daniela Carrara (1955–2014)

Contents

PART 2
The Exhibition in Detail

Preface

In 1993, the Institute and Museum of History of Science of Florence (now Museo Galileo) announced the realization of a standardized layout for cataloguing historical scientific collections. The layout, called STS (*Strumenti Scientifici*), arrived at the end of a process started in 1987, when the Museum, in accord with the Tuscany Region, began cataloguing scientific instruments of historic interest. As Mara Miniati, one of the leading members of the work group, recalled, discussions about the necessity of cataloguing standards for scientific instruments began in the early 1980s, "when it became a necessity not only to index our own holdings, but also to communicate and exchange information with other institutions."[1] Since its publication by the ICCD (*Istituto Centrale per il Catalogo e la Documentazione*), the STS became a national standard, and still is today in the revised form of PST (*Patrimonio Scientifico e Tecnologico*). The realization of the STS in 1993, however, represented not only the result of ten years' work but also the achievement of a goal outlined some 70 years before by Andrea Corsini, the very founder and first director of the Institute and Museum of History of Science.

A pioneer of History of Science in Italy, Corsini was in many respects an outsider: his interest in scientific collections, his insistence on the necessity of captivating the general public, and his practical approach towards organizational issues differentiated him from the majority of historians of science of his generation. At its earliest stage, history of science in Italy was largely dominated by history of medicine (Corsini himself was an historian of medicine). Historians of medicine could count on university chairs, institutional entrenchment, and a substantial number of followers. Being often cultivated by scientists and collectors "who lacked cognizance of the methodological and empirical problems of historiography,"[2] history of medicine rested at an amateurish level and mostly dealt with priority claims and erudite curiosities. Discussions and debates were strictly confined within the boundaries of "semi-clandestine periodicals" with no audience whatsoever outside the narrow circle of historians of science.[3]

1 Mara Miniati, "Catalogazione di strumenti scientifici: dalla scheda STS alla scheda PST," *Museologia Scientifica Memorie*, 2 (2008), 18–20: 18.

2 Maria Conforti, Historia Amabilis. *La storia della medicina nel primo Novecento*, in *Scienze e storia nell'Italia del Novecento*, edited by Claudio Pogliano (Pisa: Pisa University Press, 2007), 215–235: 227.

3 Aldo Mieli, "Organizzazioni italiane per promuovere lo studio della storia della scienza," *Archivio di storia della scienza*, 1 (1919), 94–99: 95.

Championing the idea of scientific collections as 'cultural objects' and somehow anticipating the "material turn" in the history of science,[4] Corsini proposed a comprehensive protection and promotion policy for scientific collections centered on the necessity of making the general public aware of the role of science in history and society. In order to attract the public, he maintained, history of science should look to history of art and acknowledge that staging exhibitions and establishing museums/libraries should be preferred over organizing conferences and setting up 'scientific sanctuaries' of the kind of Galileo's Tribune. This view found its most articulated expression in the First National Exhibition of History of Science held in Florence in 1929. With more than nine thousand items on display – instruments, machines, books, portraits, and memorabilia variously related to the history of science – coming from 80 Italian cities and over 200 institutions, the Exhibition was in many ways a unique event that marked the culmination point of two decades' work for the protection of the historical heritage of science. Its main goal was to offer the fullest possible overview of the Italian contribution to scientific progress; at the same time, however, it was meant to urge the Italian fascist Government to implement an effective protection policy. The Exhibition shows Corsini's amazing ability to come to terms with a complex environment, as Italy was dominated by a centralist regime which was rapidly getting mass popular consensus through its nationalism and, at the same time, shaken by subterranean centrifugal drives triggered by the Italian long-standing tradition of localism. Its unexpected success made it possible to establish in Florence a National Museum of History of Science (1930) which would soon become one of the world's leading institution in the field; the commitment it generated among a wide spectrum of Italian cultural institutions helped to put the issue of the protection and evaluation of scientific collections on the Italian Government's political agenda, and its design contributed to legitimize history of science as an academic discipline.

My aim in this essay is to examine the First Exhibition of History of Science in its cultural and political context as a case study to show how complex and ambiguous the relationships between cultural enterprises and political ideologies in 20th century Italy could be, and how the cause of science and history of science could be endorsed and preserved even underneath the layers of political propaganda spread on it at both local and national level. The idea that guided me was that, given that events such as exhibitions both match and contribute to reshape the cultural environment they are in, the First National

4 About the "material turn" see Liba Taub, "Reengaging with instruments," *Isis*, 102, 2011, pp. 689–696.

Exhibition of History of Science could offer both a privileged perspective on the Italian history of science's encomiastic and anecdotic character fed on political and ideological nationalism, and, at the same time, an insight into the progressive development of a new kind of discipline, more devoted to material and visual questions and more attentive to the general public. From this perspective, the Exhibition allows us to take a close look at the coexistence of local and national values, myths, and narratives embodied in cultural enterprises in fascist Italy, and at the mutualistic interaction between the latter and the regime, a factor of paramount importance in shaping the Italian cultural panorama at institutional level.

Even though the Exhibition failed to achieve its primary objectives – that is, special superintendencies for the history of science were never established and no general catalogue of Italian scientific relics was ever produced – it nonetheless contributed to the understanding of scientific collections and artefacts as an integral part of the Italian cultural heritage. In other words, scientific collections, once dismissed as useless and hence stashed in warehouses, began to be regarded as cultural objects. Before the Exhibition, the issue of conservation and display of scientific memorabilia had been almost exclusively addressed by a narrow circle of specialized journals and discussed in largely theoretical terms by a handful of historians of science. Thanks to the Exhibition, these issues were covered by the general press and attracted the attention of the public at large. The notion that the founding of a "National" Museum of the History of Science in Florence should be the logical – and, in a sense, expected – outcome of the Exhibition was accepted even outside the city itself and the circle of Florentine historians of science. The Exhibition therefore succeeded in demonstrating to the whole Nation the importance of protecting and showcasing the historical heritage of science. At the same time, it conveyed the idea, reiterated by Corsini, that Florence had a central role to play in this enterprise.[5] In his speech at the Exhibition closing ceremony, Ginori Conti explained that "To create this National Museum, we do not need, in Florence, to divest University laboratories and civic museums of the relics they so jealously preserve. To us, the instruments that Florence already possesses suffice, as well as those owned by private citizens. (…) To implement this program, we shall be better off if we are not deprived of help from the Authorities and especially from the Municipality, which by doing its utmost to help us will have the pride of fostering the emergence in our city (which is and must remain one of the most important centres of study and culture) a

5 A. Corsini, "Firenze e la storia delle scienze," *Atti della Società Colombaria di Firenze dell'anno 1925–1926* (Florence: 1927), 5 (offprint).

Museum unique in its kind in our Nation, which, eventually, can be a subject of pride for Florence."[6]

This outcome was the result of the idea that the Galilean tradition basically summed up the Italian tradition. The Museum that was established right after the Exhibition housed the collections previously on display in Galileo's Tribune; yet they were now recognized as the main elements of a national tradition and not just of a Florentine one. From this perspective, the new museum was a hybrid institution like no other in the Italian panorama: being funded almost exclusively by the City of Florence, it was a Civic Museum; since its collections were property of the University of Florence, it was a University Museum; finally, as it put on display a 'Galilean tradition' which was supposedly identical to the 'Italian tradition', it was a National Museum.

In another sense, the Exhibition showed that historical scientific collections could be the subjects of events addressed to a general public and that organizing such events ought to be the first preliminary step toward the creation of permanent expositions. The Natural History Museum of Trento, for example, turned its show into a permanent exposition, and in Naples a special section devoted to the History of Technology was established within the National Museum. This new section, inaugurated on December 3, 1932, featured the models built for the exhibition of 1929 and new ones executed for the occasion. Among them, there were hand- and horse- mill-stones, parts of the *fistulae* which used to carry water to *Calidarii*, models of olive-presses, *epistomia* (faucets), scales, and a scale model of a Roman pool. Another museum of History of Science devoted to Antonio Pacinotti was created in Pisa, and in Livorno Piemontese the City Councils installed a museum in Galileo Ferraris's house. In Padua, the Exposition ignited debates about the opportunity of creating a museum of machines dedicated to the engineer Enrico Bernardi (the museum would be eventually inaugurated in 1941). These examples show that the Exhibition of 1929 played a fundamental role in defining the Italian approach toward historical scientific collections and significantly contributed to the shaping of Italy's geography of historical scientific museums.

On the other side, if one looks at it from a broader perspective, the Exhibition, while standing out for its size, appears as part of an international movement that increasingly considered science and history of science as object of musealization.[7] In the same year 1929, for example, an International Exhibition within which science retained a prominent role was held in Barcelona, Spain.

6 "La chiusura della Mostra Nazionale di Storia della Scienza," *La Nazione*, 12 November 1929.

7 The word "musealization" comes from the German *Musealisierung* and roughly means the "institutionalization of the past" enacted when an artefact is deemed worthy of being

The conviction that museums were privileged environments to enact the idea of a national identity and thus constituted an effective instrument of mass education was quite common at the turn of the century and was further juiced up by the rise of local nationalisms. Paired with the positivist view of science as a progressive enterprise leading humanity from a state of superstitious ignorance up to the ethereal regions of truth, this conviction eventually led to the creation of institutions like the Deutsches Museum (1903), whose main aim was displaying and popularizing science.

According to the original plan, the Exhibition was supposed to be structured into five disciplinary sections: 1. Natural Sciences; 2. Medicine and Pharmacy; 3. Mathematics, Physics, and Chemistry; 4. Astronomy and Geography; 5. Technology. This arrangement, however, conflicted with the long-standing Italian tradition of localism. As soon as it was communicated to local committees, rumours began to spread that the actual goal of the Exhibition was to give life to a National Museum of History of Science and that objects sent to Florence would not be returned. Thus, many local committees decided to withdraw their participation in the event unless they were provided with appropriate guarantees. In the end, an agreement was reached: the Exhibition was to be "a collection of regional exhibitions" and participants were to choose by themselves what to put on display.

The Exhibition, then, was arranged geographically, and this made it impossible to display the historical development of individual sciences as Corsini had originally planned to do. Cities and institutions were invited to sort out items in their possessions and were allowed to design and manage their own exhibits. These resulted in a general lack of coherence: exhibits presented themselves one after another without apparent order beside the fact that they happened to come from the same city or region. Moreover, even though the organizers would have preferred to put on display original items, participants were allowed to present copies or even photographic reproductions of originals. As historian of science Giuseppe Montalenti put it, this undermined both the historical and didactical value of the Exhibition, making it just an inordinate succession of shows with no tie to each other.

On the positive side, the geographical ordering made the task of designing the show immensely simpler. Also, it implicitly invited visitors to compare the participating cities, thereby underscoring the richness of the Tuscan heritage

---

publicly displayed. I use it in the sense discussed by Sharon MacDonald in *Memorylands: Heritage and Identity in Europe Today* (New York: Routledge, 2013), 138.

and of the Florentine one in particular.[8] As the attempt to present a consistent national narrative through chronological and disciplinary arrangement proved unfeasible, the organizers took a reductionist approach by equating the "Italian tradition" with the Galilean/Florentine one. This was an idea especially favoured by Garbasso: for him, Galileo stood out as the prototype of the Italian genius and the 'Galilean tradition' basically summed up the whole Italian scientific tradition. Every great advancement in science, according to him, could be traced back to the 'experimental method' envisaged by Florentine physicians like Taddeo Alderotti in the Middle Ages, enhanced by Leonardo during the Renaissance, and fully developed by Galileo in the seventeenth century. If the Exhibition was to be a collection of local traditions, then the Florentine one could be presented as the main tradition whence others had grown like flowers from a stem.

The internal structure of the *Palazzo delle Esposizioni* – a large building located at the edge of the city center, which served as seat for many Florentine exhibition until its demolition at the end of 1930s – comprising two large central halls on the ground floor that lead into smaller rooms all around, favoured this type of approach and was ably exploited by the organizers. The two large halls were, in fact, reserved for Tuscany and Florence. These focused on Leonardo and Galileo, respectively, as the ideal representatives of the Italian scientific genius and founding fathers of the national tradition. Significantly, to prepare the Tuscan room, the organizers not only combined the individual exhibits of ten cities, but also broke their own rule of geographical arrangement. At the center of the hall stood a *Tribuna di Leonardo* prepared with contributions from the Biblioteca Ambrosiana in Milan and the Naples committee.[9] The centerpieces of the Leonardo section were the reconstructions of parts of the "flying machine" designed by the illustrious Tuscan. This was a project actively promoted by Corsini. In 1928, he had contacted Raffaele Giacomelli,[10] a consultant to the Ministry of Aeronautics, chief editor of the magazine *L'Aerotecnica*, librarian of the Istituto Sperimentale di Aeronautica in Rome, and – since 1926 – author of research studies on Leonardo's codexes

8 On this topic, see G. Montalenti, "La Prima Esposizione Nazionale di Storia della Scienza a Firenze," *Archeion* 11 (1929): 239–241 (Montalenti had curated the Rome room), and S. Timpanaro, "Nel mondo della scienza: alla Mostra scientifica di Firenze," *Italia letteraria*, 2 June 1929.

9 The Naples commission, chaired by Roberto Marcolongo, had prepared a replica of the instrument that Leonardo had used to solve "Alhazen's problem" concerning reflection. The replica was displayed in the Tuscany room.

10 R. Giacomelli, "I modelli delle macchine volanti di Leonardo da Vinci," *L'Ingegnere* 5 (1931): 3 (offprint).

on flight later published in the volume *Gli scritti di Leonardo da Vinci sul volo*.[11] Giacomelli pulled out a number of indications from the *Codex Atlanticus*, the *Codex B*, and the *Codex on the flight of birds*. He forwarded them to Giuseppe Schneider, a draftsman at the *Stabilimento di Costruzioni Aeronautiche* in Rome (an aeronautical construction facility), who prepared the blueprints. The models were built at the *Stabilimento* in Rome and the *Istituto Tecnico Leonardo da Vinci* in Florence headed by Alberto Picchi, who assigned the Institute's master craftsmen Mario Bucci and Vasco Menici to the task.[12] When the Exhibition closed, the models built in Florence stayed in the Museum of the History of Science, while those made in Rome were donated to the Science Museum in London. Schneider's blueprints, together with photographs of the models and Leonardo's sketches, were displayed at the Aeronautical Exposition in Milan in 1934, thus inaugurating a fruitful practice of exchanging and collaboration between institutions, and contributing to establish the Florence museum as a leading international centre for the study of scientific objects.[13]

This book is divided in two parts. In the first one, I trace the history of the Exhibition, its cultural and material premises, its organization and setup, its goals and its results. In doing this, I will highlight the local and national factors, and the ambitions and perspectives which shaped the Exhibition's ideological profile both at cultural and political level. The second part, on the other side, contains a room-by-room description of the Exhibition. In it, the reader will find information about each local committee, in a who-did-what fashion. This part is primarily intended as a reference tool, but it can also show the Italian very specific – and, under many respects, unique – context. The link between exhibitions, museums, and the rise of national narratives has been thoroughly studied by scholars, and it became apparent that the Italian reductionist approach stands apart in the European context. While France and other European countries tried to build their national narrative by establishing a clear distinction between regional and national museums, and thus between first- and second-class artefacts, this never happened in Italy. Usually, the Italian State simply took over what already existed and renamed it. So, the 'new' National Museums, being grounded as they were on local narratives,

11 R. Giacomelli, ed., *Gli scritti di Leonardo da Vinci sul volo* (Rome: 1936); on the machines in the Florence Exposition, see p. 95.

12 See A. Poggi, "Le opere immortali di Leonardo e di Galileo alla Mostra Nazionale di Storia delle Scienze," *La Nazione*, 7 May 1929; D. Brogi, "La macchina volante di Leonardo ricostruita alla Mostra della Scienza," *Il Giornale d'Italia*, 26 June 1929; I. Del Giudice, "La ricostruzione della macchina volante di Leonardo," *Unità Cattolica*, 28 July 1929.

13 R. Giacomelli, "Progetti vinciani di macchine volanti all'Esposizione Aeronautica di Milano," *L'Aerotecnica* 14 (1934): 8–9 (offprint).

weren't really 'national'.[14] The Exhibition of 1929 was no exception: it started out as an attempt to build a national narrative and ended up – for reasons that will become clear in the following chapters – as a collection of local exhibitions. Even its unifying theme, the "Italian genius," was dismembered: there was no Italian genius, but rather a crowd of 'Tuscan', 'Milanese' etc. geniuses. In this sense, the first part of the present book can be considered a case study of how such reductionist approach worked, and the second a survey of the results it yielded.

14 See S. Berger, "National Museums in between Nationalism, Imperialism and Regionalism, 1750–1914" in P. Aronsson, G. Elgenius (eds.), *National Museums and Nation-Building in Europe, 1750–2010* (New York: Routledge, 2014): 13–32; and S. Troilo, "National Museums in Italy: A Matter of Multifaceted Identity," in P. Aronsson, G. Elgenius (eds.), *Building National Museums in Europe 1750–2010*, EuNaMus Report No 1, 2011: 461–496.

Acknowledgements

This book is the result of a research project spread over several years and funded largely by the PhD. Program in History of Science of the University of Pisa with the support of the Museo Galileo in Florence. It would not have been possible without the museum staff's constant assistance. I wish to thank the then Director, Paolo Galluzzi, and vice-director, Marco Beretta, for their enthusiasm in the project and the fruitful conversations they entertained with me about it; Alessandra Lenzi, head of the Museum Library, and the late Daniela Carrara, head of the Museum Archive, for the uncountable times they fulfilled my requests; all the museum staff and consultants: Andrea Bernardoni, Sara Bonechi, Stefano Casati, Giovanni Di Pasquale, Elisa Di Renzo, Francesca Fares, Gianna Gheri, Laura Manetti, Iolanda Rolfo, Camilla Rotoli, Giorgio Strano, to name just a few. To them and to all the others I did not mention by name goes out my gratitude. I also wish to thank Massimo Bucciantini, Michele Camerota, Franco Giudice, Patrizia Ruffo, and the editorial board of *Galilaeana*. The editorial staff at Brill has been wonderful in enduring my writing and rewriting, and so has been my lovely family: my wife Elisa and my twin daughters Adele and Anna. Thank you for bearing me all these years.

Illustrations

Charts

Abbreviations

ArchMG	Archivio del Museo Galileo
ASCFi	Archivio storico del Comune di Firenze
ASDFAA	Archivio storico del dipartimento di Fisica e Astronomia dell'Università di Firenze, Arcetri
ASF	Archivio di Stato di Firenze
ASUFi	Archivio storico dell'Università di Firenze
DBI	Dizionario Biografico degli Italiani

PART 1

The Exhibition in Context

CHAPTER 1

To Protect and to Promote

1 The Italian Scientific Heritage at the Turn of the Century

On September 21, 1922, Andrea Corsini, a 47-years old Florence-born physician and historian, was attending in Bologna the conference of the *Società di storia critica delle scienze mediche e naturali* (Critic History of Medical and Natural Sciences Society), a society he was a member of since its foundation in 1909. The Society used to meet once a year to discuss themes related to the history of medicine, and Corsini himself had contributed papers on several occasions. That year, however, he was not going to deliver the habitual speech about some physician of the past or the importance of the history of medicine, for he wished, instead, to call the attention of his colleagues to a different subject, an issue he had been dealing with for the last ten years: the disgraceful condition of historical scientific collections.

The son of a manager of the Habsburg-Lorraine Tuscan Estates Directorate, Corsini had graduated in medicine at the University of Florence in 1899 and then went on to become assistant to physician Giorgio Roster at the Institute of Hygiene in Florence. While working with Roster, he became interested in the history of medicine and planned to write a history of epidemics occurred in Florence from the Middle Ages to the 20th Century.[1] The project was never realized because in 1911 he was almost accidentally led to the study of historical scientific collections when, in prevision of the International Exhibition of Hygiene which was to be held in Rome the following year, he was charged with the task of gathering information about historical scientific collections existing in Florence. Impressed by their richness and variety, he continued pursuing his research after the Exhibition closed, reaching out to librarians, archivists, and museum curators to ask them about the location, composition, and condition of collections they happened to know something about. What he found out was discouraging: uncared-for instruments laying heaped up in depots, manuscripts haphazardly stashed in warehouses, mouse-nibbled books bound to end up in pulpers, catalogues and inventories nowhere to be found, and personnel who hardly knew what they were talking about. "Whoever has the chance of combing through scientific and bibliographic stuff," Corsini remarked at very

1 For a full biography, see the entry in *Dizionario Biografico degli Italiani* (hereafter cited as DBI). The *Dizionario* can be found at the following link: http://www.treccani.it/biografie/.

 | DOI:10.1163/9789004297418_002

beginning of his speech in Bologna, "can see that a large part of it is doomed to ruin and dispersion as it lies untidy, neglected, and unsupervised."[2]

Corsini's words echoed those of historian and mathematician Guglielmo Libri, who, nearly a hundred years before Corsini, had denounced the same situation in the *avertissement* to his *Histoire de sciences mathématiques en Italie* (1838). "The immense losses already caused by our fathers' negligence have made me foresee those still to come," Libri lamented. "Our indifference towards the works that contributed the most to the advance of Reason distressed me very much."[3] It seemed like nothing had changed, despite the political, social, and cultural upheavals occurred in the meanwhile.

The protection and evaluation of the artistic and cultural heritage had been one of the many issues the earliest Italian Government had to focus on. In 1861, the newly-instituted Italian State commissioned art historians Giovanni Morelli and Giovanni Battista Cavalcaselle to prepare an inventory of the artifacts held in Umbria and Marche that the Kingdom of Italy had acquired from the suppressed religious orders by virtue of decree. In their accompanying report, Morelli and Cavalcaselle urged the Government to strengthen measures to protect cultural artifacts against deterioration, art theft, and illegal export. They suggested, as a first step, to prepare inventories as complete as possible.[4] Two years later, claiming that it was necessary to intervene as soon as possible, Cavalcaselle directly addressed the Government. In a memorandum outlining a feasible and effective cultural policy, he proposed the establishment of a governmental authority to supervise and train cataloguing staff, to coordinate restoration activity, and to guard against illegal exportation of artifacts. "It is on the Government," Cavalcaselle admonished, "that falls the task of preserving the glorious artistic tradition of the Country."[5]

After the formation of an advisory board (1867), in 1875 the Government established a directorate-general within the Ministry of Public Education and charged it to compile a general catalogue. This was a very compelling task, for in 1866 the Italian state had acquired many artifacts previously held by

2 Andrea Corsini, "Per il patrimonio storico-scientifico italiano," *Archivio di storia della scienza* 5 (1924): 348.

3 Guglielmo Libri, *Histoire des sciences mathématiques en Italie* (Paris: Chez Jules Renouard et Cie, 1838), x.

4 See Donata Levi, *Cavalcaselle. Il pionere della conservazione dell'Arte* (Torino: Einaudi, 1988), and Donata Levi, "Il viaggio di Morelli e di Cavalcaselle nelle Marche e nell'Umbria," in *Giovanni Morelli e la cultura dei conoscitori*, ed. Giacomo Agosti et al. (Bergamo: Lubrina, 1993).

5 Giovanni Battista Cavalcaselle, *Sulla conservazione dei monumenti ed oggetti di belle arti e sulla riforma dell'insegnamento accademico* (Torino: Tipografia La Subalpina, 1863), 5.

religious orders but some of the local institutions which were in charge of their preservation were in fact selling them for budgetary reasons. The directorate (first called *Direzione generale centrale degli scavi e musei del Regno* and then, from 1881, *Direzione generale per le antichità e le belle arti*) opened regional offices in 1891, and from these regional offices the so-called *Sovrintendenze* (or *Soprintendenze*) followed in 1904. Three years later, the *Sovrintendenze* were defined by law, and finally, in 1909, the so-called "Rosadi Act" decreed what ought to be subject to their care and surveillance. According to the Rosadi Act, the *Sovrintendenze* should oversee everything, mobile and immobile, "of historic, archaeological, palethnological, or artistic importance." Also, the act imposed some restrictions on private property, conferring pre-emption rights to the Italian State and, under certain conditions, the right to expropriate. Export was forbidden unless explicitly authorized by the *Sovrintendenze*, which were required to maintain an up-to-date catalogue of objects, buildings and artifacts.[6]

Such measures had had modest or no effect at all on historical scientific collections. As Corsini made clear in his speech, "while for everything pertaining to Letters and Fine Arts there is always someone in charge of overseeing and preparing catalogues, no one minds about scientific relics."[7] Appointed officials of the *Sovrintendenze*, he lamented, were either art historians or archaeologists who had little, if any, interest in science. Moreover, history of science itself was not fully acknowledged as an academic discipline. Voluntary classes in history of science, usually held by scientists, had been suppressed in 1911, and historians of science were often regarded as 'wannabe' philosophers by humanists and as 'failed' hard-science researchers by scientists. "History of science is held in dubious esteem by those who work for the progress of science," Federigo Enriques would later affirm. "For the most part, it interests only aged scientists who, as they grow weary of studying new things, complacently turn their attention to researching the nobility titles of their discoveries."[8] Enriques, a trained mathematician, during the 1910s had championed the cause of a 'scientific philosophy' within which history of science should have been the necessary bridge between humanistic and scientific culture, but met with harsh criticism from philosophers. Benedetto Croce, who considered

6 See Andrea Emiliani, "Musei e Museologia," in *Storia d'Italia, i documenti, V* (Torino: Einaudi, 1973), 1615–1655.

7 Corsini, "Per il patrimonio storico-scientifico," 351.

8 Federigo Enriques, *Il significato della storia del pensiero scientifico* (Manduria: Barbieri, 2005), 11.

scientific notions as mere 'pseudo-concepts', publicly exhorted him to get back to mathematics and leave philosophy and history to real philosophers.[9]

Corsini, however, had a strong disposition towards practical thinking and was barely interested in this kind of arguments. He was a brilliant and capable organizer, gifted at pinpointing factual problems, and skilled in planning for both short- and long-term goals. Firmly believing that science and history of science deserved to be recognized as important constituents of Italian culture alongside art and literature, he demanded fair treatment from the Ministry in terms of funding and trained personnel rather than 'philosophical acknowledgment'.

> Whoever guides you through bibliographical collections will certainly let you know if there are rare editions of Dante, Petrarca, and Virgil, or letters by Raffaello Sanzio and Garibaldi, but you will learn nothing about science books, since they are generally considered old and useless stuff. There are scientists who wrote huge notebooks and kept massive correspondences: is there someone who knows them? – It is not, for scientists, as it is for artists, among whom even secondary figures are renowned; in science, either you are a Galileo or you are nothing.[10] […] There is no doubt that it is necessary to preserve science collections, and that, just like we spend so much money for the preservation of fine art, we should also spend a little money – I repeat: a little money – for this. Otherwise, let's have the courage to auction or sell the collections themselves to the Americans: that would be better than seeing them in ruin. At least, we would secure them and earn some money![11]

The provisions proposed by Corsini reflected his sound practical sense. They were intended to "save and capitalize on this asset [i.e., scientific collections], increasing the nation's credit," and were modeled after what had already been done for fine arts. First of all, it was necessary to get as clear an idea as possible of the extent of the collections by preparing catalogues and inventories; then, keen and competent supervisors should be appointed; finally, small exhibitions and museums must be established. As for a possible library of history of science to support the supervisors' and scholars' work, Corsini suggested a handy solution. "I am not going as far as to ask the same of what Mieli wishes,

9 On these topics, see Michele Ciliberto, "Scienza, filosofia e politica: Federigo Enriques e il neoidealismo italiano," *Studi Storici* 22 (1981), 861–886.

10 Corsini, "Per il patrimonio storico-scientifico," 350.

11 Corsini, "Per il patrimonio storico-scientifico," 354.

that is, a library exclusively devoted to the history of science," he stated, "for I would content myself if catalogues in major libraries featured history of science as a subject."[12]

It was precisely this kind of approach that distinguished Corsini from scholars like the aforementioned Aldo Mieli, with whom he was on very amicable terms notwithstanding. Mieli had started his career as chemist, but attained his greatest reputation as a historian of science. In 1919, he had founded the *Archivio di storia della scienza*, which would soon become the leading Italian journal in the field. Strongly influenced by Ernst Mach, Wolfgang Ostwald, and George Sarton, Mieli grounded his ideas about history of science on the claim that there was no split between science and philosophy, for they both were part of the same intellectual activity aimed at a systematic knowledge of reality. Therefore, history of science, being the study of the progressive development of such systematic knowledge, should be kept apart from the history of individual sciences.[13] This view eventually resulted in an emphasized centralism in both planning historical researches and managing historical scientific collections – the major tasks a historian of science, according to Mieli, should attend to. In an open letter addressed to the Minister of Public Education, Giovanni Gentile, in September 1923, Mieli endorsed the creation of a National Library and Museum of History of Science in Rome to gather and house books, manuscripts, and artifacts from all over Italy.[14] For him, creating a museum was the necessary condition for the institutional and cultural recognition of history of science, and such museum ought to be 'national' so that its very existence could constitute what contemporary museum studies call "a performative utterance of having an identity."[15]

It was an ambitious project, and yet, given the peculiar Italian situation, hardly feasible, at least in the short run. Just like fine art collections, also historical scientific ones had followed the fate of Italy's decentralized past. Still, Mieli was reluctant to recognize that in order to achieve his goal a complete

12 Corsini, "Per il patrimonio storico-scientifico," 354.

13 On Aldo Mieli, see Claudio Pogliano, "Aldo Mieli, storico della scienza," *Belfagor* 28 (1983), 537–557; Massimo Bucciantini, "George Sarton e Aldo Mieli: bibliografia e storia della scienza a confronto," *Nuncius* 2 (1987), 229–239; Ferdinando Abbri, "Aldo Mieli: la storia della scienza tra sviluppo del pensiero e pratica scientifica," in *Scienze e storia nell'Italia del Novecento*, ed. Claudio Pogliano (Pisa: Plus, 2007), 49–66.

14 Aldo Mieli, "Per una biblioteca di storia della scienza: lettera aperta a Giovanni Gentile," *Archivio di storia della scienza* 4 (1923), 305–306.

15 Sharon MacDonald, "Museums, national, postnational and transcultural identities," *Museum and Society* 1 (1), 2003, pp. 1–16: 3.

reshaping of the Italian cultural geography, especially at the institutional level, was required. Corsini, on the contrary, was perfectly aware of the situation:

> Obviously, there is no need to move, transfer, revolutionize. Collections can stay where they are, but on the condition that there will always be someone to take care of them. After all, it is not like paintings and other art objects are all kept in our Galleries! They are also in churches, monasteries, etc.; but, even there, no one can harm them, let alone export them without governmental permit. So, let us decide what to do on a case-by-case basis. Even though I am not saying to remove collections from their place, yet I say that surely it would be possible to keep an eye on how they are managed. There are directors and professors who maintain the collections rightly as it ought to be, and also try to improve, organize, and augment them; but, on the other hand, there are directors who have left everything go to ruins. We absolutely must not permit that. [...] No one would get offended, I guess, if we undertook a monitoring, as this is already mandatory for important museums and libraries. In all regions across Italy there are officials guarding over monuments: they are educated and, above all, enthusiastic, and they cost nothing to the State, except a few money for travel and mail reimbursements, since their appointment is honorary. I know for a fact that these officials have done and are still doing a very good and sometimes even invaluable service to the State. Why not appoint similar inspectors for scientific collections as well?[16]

Corsini knew that local museums, no matter how small or unknown, were quite unwilling to hand their collections over to a National Museum of the kind Mieli envisioned. To achieve practical results, it was necessary to reckon with the peculiar Italian decentered context, not try to wipe out everything just to imitate what had been done in Germany or France.

2 Universities, Museums, and the History of Science

At the turn of the XIXth century, numerous and valuable historical scientific collections were scattered among public and private Italian institutions. Collections of anatomy and natural history, museums of physics, astronomy, and medicine were often housed in civic museums and Universities, and

16 Corsini, "Per il patrimonio storico-scientifico," 351–352.

sometimes existed as independent institutions. Many had been established thanks to bequests from renowned scientists, and after those same scientists, just like tutelar deity, they were also named. Generally, they had been created for the purpose of gathering scientific instruments no longer in use and to preserve books and manuscripts belonged to famed scientists. Constantly struggling with lack of space and insufficient funding, they often consisted of a couple of rooms and had no staff other than occasional volunteers. Rarely their collections were indexed, and seldom, if ever, they were open to the public. In the end, passion and tenacity of their directors were the only asset they could rely on.[17] Such were the cases, for example, of the *Museo Aldrovandiano* of Bologna, established thanks to a bequest from Ulisse Aldrovandi, and of the *Museo Torricelliano* of Faenza, which was the result of the *Esposizione Universale Torricelliana* of 1908.

A passionate collector of natural curiosities, during the course of his life Aldrovandi had accumulated in his house a huge collection made up of more than eighteen thousand "varieties of natural things" and a fifteen-volume herbarium featuring over seven thousand dried plants. Before dying in 1605, he had left this extraordinary "microcosm of nature" to the Senate of Bologna with the purpose of creating a public museum to increase the glory of the city. The entire collection was housed in the Palazzo Comunale until 1742, and then transferred to the Science Institute at Palazzo Poggi together with another scientific collection donated to the City by marquis Ferdinando Cospi in 1657. Once at Palazzo Poggi, part of the Aldrovandi's collection joined the *Museum Diluvianum* and the natural history collection set up by Giuseppe Monti and Luigi Ferdinando Marsili respectively, while the rest was dismembered and distributed among several Bolognese institutions. It was not until 1907 that the *Museo Aldrovandiano* was partially reunited in Palazzo Poggi again.[18]

The history of the *Museo Torricelliano* began in 1864, when the city of Faenza erected a statue of Evangelista Torricelli with the explicit aim to lay claim to being Torricelli's homeland against Imola.[19] Forty-four years later, on occasion of the sixth centenary of Torricelli's birth, Faenza hosted an exhibition called *Esposizione Universale Torricelliana*, within which the figure of Torricelli was

17 For an overview of Italian scientific museums and collections at the turn of the century, see Elena Canadelli, "I musei scientifici," in *Storia d'Italia. Annali 26. Scienze e cultura dell'Italia unita*, ed. Claudio Pogliano and Francesco Cassata (Torino: Einaudi, 2011), 867–893.

18 See *Guida ai Musei di Palazzo Poggi: scienza e arte* (Bologna: Compositori, 2001), ed. Walter Tega.

19 See Giovanni Ghinassi, *Per l'innalzamento della statua di Evangelista Torricelli in Faenza nel CCLVI della sua nascita* (Faenza: dalla Tipografia di Pietro Conti, 1864).

celebrated as part of the Faentinian cultural and artistic heritage. After the exhibition closed, the collections on display were transferred to the local art gallery in prevision of establishing a permanent museum. They remained there until 1920 and were afterward moved to the Civic Library of Faenza, where still there were no room to set up a proper museum. Only in 1923 the Library Director, Piero Zama, managed to establish it in a tiny space.[20]

Public museums entirely devoted to the history of science like the *Museo Aldrovandiano* and the *Museo Torricelliano* were rare. In fact, most of the historical scientific collections were owned by Universities. Every Italian University city had its own historical scientific collection, be it noteworthy and well-preserved or forgotten and unknown even to locals and University staff. The *Teatro di Filosofia sperimentale* of the University of Padova, founded in 1740 by Giovanni Poleni, had been the very first physics cabinet in an Italian University. Himself an impassioned philologist and antiquarian, Poleni had furnished the *Teatro* not only with research and didactic tools, but also with Sixteenth- and Seventeenth-century scientific instruments. Still, in the 20th-century this museum was barely known by general public and received little institutional support.[21] On the other hand of the spectrum lay the rich and well-supported physics cabinet of the University of Pavia. Established in 1771 in the context of the higher education reform in Lombardy promoted by archduchess Maria Theresa, it was led since 1778 by Alessandro Volta. It was equipped with high-end instrumentation that, after Volta's death in 1827, was gathered together to constitute a museum which, at the end of the 19th-century, had almost two thousand objects on display. Unfortunately, a large part of Volta's collection burned to ash in a disastrous fire that broke out at the Voltian Exhibition in Como, 1899.[22]

In Rome, the *Museo copernicano* and the anatomical collection donated by Giuseppe Flajani to the University of Rome in 1870 suffered different fates. The *Museo Copernicano* had been established in 1882 on the initiative of the University of Rome and thanks to the bequest of Polish collector Artur Wołyński. The museum featured astronomical instruments and copernican

20 On the *Esposizione Torricelliana*, see *L'Esposizione di Faenza del 1908*, eds. Claudio Casadio and Anna Rosa Gentilini (Faenza: Casanova, 2008).

21 On the *Teatro*, see *Il teatro di Filosofia sperimentale di Giovanni Poleni*, eds. Gian Antonio Salandin and Maria Pancino (Trieste: Lint, 1987); Gian Antonio Salandin, "Dal Teatro di filosofia sperimentale di Giovanni Poleni al Museo di storia della fisica," in *I musei, le collezioni scientifiche e le sezioni antiche delle biblioteche*, ed. Carlo Gregolin (Padova: Edizioni dell'Università di Padova, 1996), 101–108.

22 See *Gli strumenti di Alessandro Volta: il Gabinetto di fisica dell'Università di Pavia*, ed. Giuliano Bellodi (Milano: Hoepli, 2002); Alessandra Ferraresi, "Il Gabinetto di Fisica sperimentale nella seconda metà del secolo XVIII: didattica, divulgazione, ricerca nella politica asburgica della scienza," in *Annali delle Università Italiane* 7 (2003), 91–110.

relics, and was housed at the Observatory of the Roman College. In 1923, When the Observatory was annexed to the Central Office for Meteorology to create the Royal Observatory of Rome, the *Museo Copernicano* was dismantled and its collections provisionally transferred to the Lincei Academy. They would be kept there until 1932, virtually unvisitable,[23] just like virtually unvisitable was also, at the beginning of the 20th-century, the anatomical collection gathered by Flajani and stashed in the warehouse of the *Santo Spirito* Hospital.[24]

University warehouses were literally filled with instruments no longer in use. The *Regia Scuola di Applicazione per gli Ingegneri* of Pisa, founded in 1913, kept electromagnetic machines, manuscripts, and drawings by Antonio Pacinotti; a huge collection of scientific instruments was stored at the University of Urbino; other valuable collections were owned by Universities in Turin, Siena, Parma, Brescia, Bergamo, Genova, Catania, Cagliari, Modena, Naples, Palermo, and so on.[25] Still, more than actual museums, these places were just refurbished warehouses. Having no tie whatsoever with libraries and archives, they carried out no research activity. Their main aim seemed to celebrate the glorious and lavish past of the institutions that housed them: in a way, they were kind of silent and austere scientific reliquaries.

In Florence, the *Tribuna di Galileo* and the *Museo degli Strumenti Antichi*, both part of the *Imperial Regio Museo di Fisica e Storia Naturale* founded in 1775 on the initiative of Grand Duke Pietro Leopoldo, testified to the Medicean and Lorrainese traditions of collectionism. The first project of the *Tribuna* dated from 1829, but the construction had been completed and inaugurated in 1841, on the occasion of the Third Conference of the Italian Scientists. Conceived by the Grand Duke Leopoldo II and the director of the *Regio Museo*, Vincenzo Antinori, as a solemn celebration of Galileo's genius, the *Tribuna* had been designed by architect Giuseppe Martelli in a severe neoclassical style resembling that of a church. Perhaps more befitting a Christian Saint than a secular man, this "scientific sanctuary" (as Antinori called it) commemorated Galileo in true hagiographical fashion. Under the dome, at the center of the hall, a marble statue of Galileo stood, surrounded by niches filled with the busts of Galileo's disciples, a few of his instruments, and the middle finger of his right hand. On the walls, several frescoes depicting scenes from Galileo's

---

23 See Marinella Calisi, *Storia e strumenti del Museo astronomico e copernicano di Roma: guida alle collezioni* (Roma: Osservatorio astronomico di Roma, 2000).

24 See Tina Bovi, *Il Museo Storico Nazionale dell'Arte Sanitaria* (Roma: Fratelli Palombi Editori, 2001).

25 More information about these and other museums and collections can be found in the second part of this book.

life celebrated him as the father of experimental science.[26] The *Museo degli Strumenti antichi*, with its four rooms devoted to Antonio Santucci, Giovanni Battista Amici, Leopoldo Nobili, and Alessandro Volta, although looking rather low-key as compared to the *Tribuna*, was animated by the same commendatory spirit.[27]

Given such celebratory purpose, the *Tribuna* and the *Museo degli Strumenti Antichi* remained in their seat at Palazzo Torrigiani when most of the scientific collections of the *Regio Museo* were transferred to the newly instituted *Regio Istituto di Studi Superiori e Pratici* to serve as didactic tools. Once suppressed the *Regio Museo*, then, the number of visitors dropped dramatically. "By now," Corsini wrote in 1924, "Florentines have sent them to the scrapheap. As for foreigners, they do not even think of the existence of science museums besides Art Galleries, since tourist guides –who continuously swarm about the Uffizi Gallery, the Pitti Palace, the San Marco Square, and even the now restored *Palagio di Parte Guelfa*– do not let them know about them."[28] Obviously, Corsini was well aware that neither tourist guides nor the public itself could be blamed for the situation. The public, he maintained, needs to be engaged, but scientists and scholars seemed barely interested in doing that. "Since when scientists themselves got away from the rooms housing the scientific collections to hole up in laboratories," Corsini complained, "scientific museums, which nobody hears nothing about anymore, are empty and depleted."[29] To engage public, Corsini thought, it was necessary to blur the distinction between research, teaching, and popularization. "Scholars must get out of their ivory tower and go deal with the public," he urged. Bring history of science to the public, and the public will repay you.[30]

26 On the *Regio Museo*, see Simone Contardi, *La casa di Salomone a Firenze. L'imperiale e reale Museo di fisica e storia naturale* (Firenze: Leo S. Olschki, 2002). On the Tribuna: Fausto Barbagli, "«Il Tempio ove sorge siccome nume». La Tribuna," in *Il Museo di Storia Naturale dell'Università degli Studi di Firenze. Volume I. Le collezioni della Specola: Zoologia e cere anatomiche*, eds. Giulio Barsanti and Guido Chelazzi (Firenze: Firenze University Press, 2009), 37–56.

27 Antonio Boffito, *Strumenti della scienza e scienza degli strumenti. Con l'illustrazione della tribuna di Galileo* (Firenze: Barbèra, 1929), 209.

28 Andrea Corsini, "La Specola di Firenze," in *la Lettura* XIV, 5 (1924), 377.

29 Andrea Corsini, "La Specola di Firenze," 377–378.

30 Andrea Corsini, "La Specola di Firenze," 378.

3 Bringing History of Science to the People: Exhibitions and Commemorations

Bridging the gap between the broader public and the community of scholars had been one of Corsini's main aims since very early in his career. While still a student at the University of Florence, he had founded the *Associazione Pro Cultura*, an organisation concerned with the promotion of scientific and literary education among lower classes by means of conferences, evening courses, and small exhibitions. Later, as an activist of both the Special Commission for the Prevention of Alcoholism and the National Federation for the Fight against Tuberculosis, he had written pamphlets and had joined educational campaigns directed at common people. Finally, once engaged in the history of science, he strongly championed the popularization of the discipline. In 1912, on the occasion of the First meeting of the *Società di storia critica*, he presented a paper titled "On the Importance and Methods of Spreading Knowledge of the History of Science in General, and of History of Medicine in Particular," in which he stated his ideas about the need to popularize history of science. "We must persuade ourselves that, at the moment, our main task is to publicize and spread the word," he remarked. "Drawing attention to our studies by reaching a wider audience is the only way to assert ourselves to the Government."[31] According to Corsini, history of science could benefit from the restored interest in science determined by the ever-increasing popularity of Universal Expositions, and from the wave of nationalism sweeping through the whole of Italy.

Universal Expositions had featured retrospective exhibitions since the Great Exhibition of 1851. Representing human progress and giving substance to the idea of "technological innovation" behind it could not prescind from history, and so retrospectives were given more and more room as the ideology of progress gradually unravelled itself. In 1855, a special section of the Universal Expositions of Paris had been devoted to scientists like Joseph Priestley, Henry Cavendish, Antoine-Laurent de Lavoisier, and Alessandro Volta.[32] Relatively large exhibitions on labor history, history of inventions, and history of craftsmanship had been staged at the Universal Expositions of Paris (1867) and Wien (1873), and a retrospective about the history of transportation, curated by Achille Bertarelli, featured at the International Exposition of Milan

31 Andrea Corsini, "Relazione sull'importanza e i metodi di divulgare la conoscenza della storia delle scienze in genere e della medicina in ispecie," in *Atti del I Congresso nazionale della Società italiana di storia critica delle scienze mediche e naturali, Roma, 1912* (Grottaferrata: Tip. italo-orientale S. Nilo, 1913), 12 (extract).

32 Ken G. Beauchamp, *Exhibiting Electricity* (London: The Institution of Electrical Engineers, 1997), 116.

in 1906. The whole Paris Exposition of 1900, being an appraisal of a century of scientific and technological progress, was marked with history. As General Commissioner Alfred Picard wrote,

> Universal Expositions are, under this respect, very fruitful enterprises, especially when they include, as was the case in 1900, centennial museums and retrospectives. They are amazing education tools, since they show the continuous evolution of the world in an impressive tableau; they elevate the soul, inspire people and nations, lay the path to the future, and powerfully serve the nations' commercial and industrial interests.[33]

For countries like Italy, retrospectives were occasions to conceal, at least in part, their technological underdevelopment. "Is there someone who lives in and consoles himself with the past, and is also so rightly proud of it, more than Italians?" a statement of the Italian Committee at the Paris Exposition of 1867 asked. "And so, what better occasion than the Exhibition of Paris to sport our grandfathers' clothes and disguise ourselves as grownups?"[34] Still, even though the Italian special commission for the Labor History Exhibition at the Exposition of Paris featured famed historians such as Pasquale Villari, Luigi Passerini, Michele Amari, Francesco Bonaini, Ercole Ricotti, and Salvatore Bongi, Italy did not look as a grownup at all. The special commission failed to get Italian museums and collectors involved, and so Italy's unassuming exhibit was eventually set up mostly thanks to antiquarian Alessandro Castellani, who personally borrowed Italian artifacts from French collectors. "It was almost nothing," the Committee disconsolately observed. "And yet, with this 'almost nothing', Italy has ranked third on an evaluation of the number of objects."[35]

Italian commissioners would often come back from Universal Expositions persuaded that Italy had to create institutions like the *South Kensington Museum* of London and the *Conservatoire National des Arts et Métiers* of Paris. Such institutions defined their identity precisely in opposition to University museum and cabinets of curiosities, for their primary purpose was to promote scientific education while fostering national identity. As director of the *Regio Museo Industriale*, Giuseppe Devincenzi, wrote in 1865, they served the whole nation, not some particular city or University.[36] The *Regio Museo Industriale*

33 Alfred Picard, *Le bilan d'un siècle*, I (Paris: Imprerie Nationale, 1906), II.

34 *L'Italia alla Esposizione Universale di Parigi del 1867. Rassegna critica descrittiva illustrata* (Paris-Firenze: n.p., 1867), XLI.

35 *L'Italia alla Esposizione Universale*, XLI.

36 Giuseppe Devincenzi, *Del Museo Industriale Italiano e del progetto di legge pel suo ordinamento. Osservazioni di G. De Vincenzi* (Torino: Tipografia di Enrico Dalmazzo, 1865), 19–20.

of Turin itself was an early, failed attempt at this. Founded by Royal Decree on November 23, 1862, and put under the administration of the Ministry of Agriculture, Industry, and Commerce, it was established in order to fill the industrial gap between Italy and the most industrialized European countries by promoting technical, technological, and scientific education. Even though according to the founding decree the *Regio Museo* was supposed to be a permanent, historical, and progressive exposition of items related to art and industries, in practice it dealt almost exclusively with technical and scientific research, and in 1906 was merged with the *Scuola di applicazione per gli ingegneri*, giving life to the *Politecnico* of Turin.

However, Italian Governments soon realized that history of science could be a helpful instrument for strengthening national sentiment and promoting scientific education, and therefore began to support local initiatives such as erection of monuments, exhibitions, and publication of books. Great scientific personalities were celebrated as exemplary figures, idealized as models of probity, and idolized as universal geniuses. Characters like Leonardo da Vinci, Andrea Cesalpino, and Galileo Galilei were celebrated as the bright and masterly minds behind human progress, alternatively admired, envied, and looted by foreigners, while too often unacknowledged and forgotten by Italians.[37] The major of such events, during the post-unitarian period, had been the *Esposizione Voltiana* held in Como in 1899, one hundred years after the invention of the battery. The Exposition covered more than fifteen thousand square meters and was entirely enclosed in a wooden structure consisting of several pavilions. At the top corners of the facade, two huge battery-shaped towers dominated the vista. The Exposition featured exhibitions of electricity, industrial machinery, fine arts, floriculture, furniture, and Italian handcraft. Regrettably, it lasted only 50 days: on the morning of July 8, a ruinous fire destroyed it almost completely in less than an hour. The surviving relics were transferred to the *Museo Civico* of Como and then, in 1927, to the *Tempio Voltiano*, a building loosely inspired by the *Tribuna di Galileo* and built that same year thanks to businessman and politician Francesco Somaini.[38] Meanwhile, other exhibitions took place

37 See, for example, Massimo Bucciantini, "Galileo e le passioni del Risorgimento," in *Il caso Galileo: una rilettura storica, filosofica, teologica. Convegno internazionale di studi. Firenze, 26–30 maggio 2009*, eds. Massimo Bucciantini, Michele Camerota, and Franco Giudice, 347–366 (Firenze: Leo S. Olschki, 2011); Marco Cini, "L'ambigua costruzione di un'identità nazionale: le celebrazioni di Galileo Galilei a Pisa (1864)," in *Galileo e Bruno nell'immaginario dei movimenti popolari tra Otto e Novecento*, ed. Franco Bertolucci, 67–84 (Pisa: BFS, 2001).

38 See *Como e l'Esposizione Voltiana, 1899* (Como: Tipografia cooperativa comense, 1899); *Fascismo e Scienza. Le celebrazioni voltiane e il Congresso Internazionale dei Fisici del 1927*,

elsewhere. Expositions of history of medicine were set up in Turin (1898) and Rome (1911) by Pietro Giacosa and Pietro Capparoni, respectively;[39] retrospectives on transportations, medicine, agriculture, and industry were held in Milan (1906) and Rome (1912); in 1909, an exhibition of Galilean documents was set up at the State archive of Florence to celebrate the publication of the twentieth and last volume of Favaro's *Edizione Nazionale delle opere di Galileo*;[40] and on the occasion of the meeting of the *Società di Storia critica delle scienze mediche e naturali* held in 1922, the University of Bologna staged an exhibition of antique science books.

What is important to underline here is that, even though scientists had been at the forefront of the *Risorgimento*, history of science museums, collections, and exhibitions contributed little to the development of a national master narrative. Local narratives kept prevailing even after the unification and despite the Government's effort. Faenza celebrated Torricelli against Imola. The *Museo Aldrovandiano* was meant to increase the glory of Bologna. The *Regio Museo Industriale* showed Turin's technological and industrial primacy. The values and myths embedded in history of science museums and collections were those of the local communities they were supposed to represent. This failed articulation between local traditions and nation-building narratives put historical scientific collections at the margin of Italy's cultural life, so that they went unrecognized both by the State and the public. At the Bologna meeting, Corsini stressed on the necessity of bringing history of science to the public first. Children, he argued, are naturally interested in science: it is scholars' fault if they drift away from it as they grow up. "We need to edit catalogues that fit for the broad public's attitude and culture." he stated. "They should be richly illustrated and should contain plenty of information to arouse the non-specialized audience's curiosity."[41] With the advent of fascism and the construction of a new national narrative based on the myth of the Italian genius, celebrations of scientists grew more and more common. The regime urged cultural institutions

---

eds. Aldo Gamba and Pierangelo Schiera (Bologna: il Mulino, 2005); Antonio Marino, *Como al tempo del decoro e dell'orgoglio. L'Esposizione Voltiana, l'incendio e la ricostruzione. Gli occhi del mondo sulla sfida di una piccola città* (Como: Alessandro Dominioni Editore, 2012).

39 On the Turin Exhibition, see Pietro Giacosa, *Magistri salernitani nondum editi: catalogo ragionato dell'Esposizione di storia della medicina aperta a Torino nel 1898* (Torino: f.lli Bocca, 1901); Pietro Giacosa, *Notizie sugli oggetti esposti alla mostra di storia della medicina* (Torino: f.lli Bocca, 1898).

40 Umberto Dorini, "La Mostra di documenti per la storia delle scienze all'Archivio di Stato fiorentino," *Bibliofilia*, 10 (1908/1909), 363–390.

41 Corsini, "Per il patrimonio storico-scientifico," 353.

to retrace the footprints of great Italian intellectuals and endorsed the idea of a 'living museum' – similar to the *Deutsches Museum* of Munich – to show Italy's lasting contribution to Western Civilization. Historians like Corsini felt that there were many opportunities for history of science, and so began to put forward large-scale projects.

CHAPTER 2

From Local to National

1 The Rise of Fascism

On the morning of September 20, 1922, while Corsini was presumably penning the final touches to the paper about the protection of scientific collections he was to read the following day in Bologna, a crowd of so-called 'blackshirts' was gathering by the hundreds in Udine to witness a ground-breaking, far more menacing speech. Located in Friuli, almost at the center of a region whose borders were still a matter of dispute between Italy and the Kingdom of Yugoslavia, Udine was one of the places where frustration following the Great War had been the most acute, thus providing a fertile ground for political movements imbued with dissatisfaction towards traditional political parties.[1] It is no wonder, then, that Benito Mussolini, head of the fascist party and political leader of the blackshirts, had chosen this city to deliver a speech that would later be remembered as a cornerstone of fascist rhetoric and politics.[2]

As Mussolini took the stage of the *Teatro Sociale* he was met with huge cheers from the audience. He opened his discourse – which would be published the following day with the title *L'azione e la dottrina fascista dinnanzi alle necessità storiche della Nazione* ("The practice and theory of fascism before the Nation's historical needs") – with somewhat of an *excusatio*. "The speech that I intend to make today," he began, "is going to be an exception to the rule that I have imposed on myself of limiting the manifestations of my eloquence, as far as I can."[3] Then, in rapid-fire succession, he pointed out the building blocks of the fascists' credo: unity, violence, discipline, nation, State, regime. On several occasions, Mussolini's speech was interrupted by ovations and applause.

Fascism had formally entered the Italian political scene the year before, when 35 members of the *fasci di combattimento* (with Mussolini himself among them) were elected to the Parliament under the flag of the *Blocchi*

1 See Mario Fabbro, *Fascismo e lotta politica in Friuli (1920–1926)* (Venezia: Marsilio, 1974).

2 Elisa Martínez Garrido, "Il primo discorso fascista di Mussolini: la traccia dannunziana," *Cuadernos de Filología Italiana*, 5, 1998, pp. 213–229: 223.

3 Benito Mussolini, "L'azione e la dottrina fascista dinnanzi alle necessità storiche della Nazione," in *Opera Omnia*, eds. Edoardo e Duilio Sismel (Firenze: la Fenice, 1956), vol. 18, pp. 411–421: 411. English translation by Bernardo Quaranta di San Severino, modified by Barbara Spackman, in Barbara Spackman, *Fascist Virilities. Rhetoric, Ideology, and Social Fantasy in Italy* (Minneapolis-London: University of Minnesota Press, 1996), p. 114.

 | DOI:10.1163/9789004297418_003

nazionali, the right-wing political coalition prompted by Giovanni Giolitti. By allowing fascists within the *Blocchi nazionali*, the 79-years old Giolitti hoped to keep fascism's extremist fringe under control and exploit the movement to weaken socialists and other left-wing parties. His plan, however, soon proved delusional. As violence kept escalating and fascist political leaders inflamed the masses by launching invectives against the Liberal State's chatty parliamentarism, fascism earned itself the favour of industrialists, conservatives and middle-class people who saw bolshevism as an incumbent threat and decadence of the State as a current reality. By 1922 fascism enjoyed widespread support from Italians and presented itself as a 'true' revolutionary movement, committed to action, discipline, and strong decision-making. "Ours is a resolutive violence," Mussolini stated in Udine. "For just forty-eight hours of methodical and warlike violence allowed us to obtain what we wouldn't have obtained in forty-eight years of vain lecturing and promoting."[4]

Approximately one month later, on October 24, a general meeting of the Fascist Party was held in Naples. Mussolini, again, addressed his followers. "We want to become the State!" he cried, igniting the crowd's enthusiasm. "They will give us the power or we will seize it by ourselves marching on Rome,"[5] he thundered later that day during a gathering of blackshirts. And so they did. On October 28, fascists from all over Italy set off to Rome in what was called the *Marcia su Roma*. Mussolini himself was not present, however, for he was staying in Milan, just in case the 'fascist revolution' turned out to be a disaster. He feared that the King Vittorio Emanuele III might sign Prime Minister Luigi Facta's decree proclaiming a state of siege and thus deploy the army, so he stayed safe in Milan, ready to flee to Switzerland. Luckily for him – and much to the surprise of many – the King did nothing of the kind: in fact, he promised to nominate Mussolini as Prime Minister upon receiving Facta's resignation. On October 30, then, the new Government was presented. It featured only three members of the fascist party, yet fascism itself definitively took root in the Italian political and parliamentary world.

Mussolini's first moves as Prime Minister were aimed, above all, at securing the support of powerful lobbies and pave the way for a dictatorial turn. Between October and December, a number of provisions reduced the tax burden of high-income earners and on November 25 the Government was given full power to reorganize from head to foot the State's administrative apparatus. Quite predictably, the parliamentary elections of 1924 saw fascists,

4 Benito Mussolini, "L'azione e la dottrina fascista dinnanzi alle necessità storiche della Nazione," p. 413.

5 Benito Mussolini, "Il discorso di Napoli," in *Opera Omnia*, vol. 18, pp. 453–460: 454, 459.

reunited under the *Lista nazionale*, win control of the Parliament. Giacomo Matteotti, a Socialist deputy, publicly denounced the intimidations, assaults, and abuse perpetrated by fascists during the election period. In a heated session of the Chamber of Deputies held on May 24, 1924, he pointed out that fascist squads threatened potential left-wing candidates, assaulted socialist's offices and headquarters, beat up participants in political meetings, tried to impede the regular constitution of electoral lists, and prevented representatives of the left-wing coalition from being present at polling places. According to Matteotti, many peasants and office employees were persuaded to vote for the *Lista nazionale* under the threat of being beaten up and lose their job, and in the end "no Italian voter has been able to express his free will," he declared amidst the fascist deputies' sonorous booing.[6]

After the session, Matteotti reportedly told his friends that he knew he had just delivered his "funeral oration."[7] On the afternoon of June 10, he was kidnapped by a group of squadrists and stabbed to death. His body was disposed in a wood outside Rome, and was not found until August 16. The Government's involvement in the murder was apparent since the early stages of the official investigation led by Mauro Del Giudice: the car used to kidnap Matteotti was found to be property of Filippo Filippelli, director of the newspaper *Corriere Italiano* and a close associate of Under Secretary of Interior Aldo Finzi. Further investigations showed that Cesare Rossi, head of Mussolini's press office and chief of the secret police, was also involved in the murder. Mussolini forced both of them to resign, but this did little to appease the outrage. The opposition parties, led by Filippo Turati, withdrew from parliamentary activity in the hope of bringing down the Government and then stage new elections; yet Mussolini withstood the public scandal and on January 3, 1925, assumed on himself, in front of the Parliament, the moral and political responsibility of the murder and of all of the fascists' past violence. Mussolini's speech inaugurated a new phase in Italian political and civil life. Fascist Government, which until then had tried to maintain at least an appearance of temperance and moderation, decisively moved towards a full and bright-clear dictatorship. Between the years 1925–1929, a number of new laws modifying the State institutional asset and severely limiting personal and collective liberties were approved. The Grand Council of Fascism was declared as the supreme organ of the State, workers' unions were suppressed, and political opposition outlawed. Death penalty was reintroduced as a punishment for "anti-Italian" activities and corporativism was embraced as the new economic policy.

6 "Il discorso di Giacomo Matteotti alla Camera dei deputati del 30 maggio 1924 (Resoconto stenografico)" in *Tempo Presente*, n. 400–402, 2014, pp. 16–30.

7 See Emilio Lussu, *Marcia su Roma e dintorni* (Milano: Mondadori, 1968), p. 155.

The rise of fascism had a significant impact on Italian cultural life. Right from the start, the regime put great emphasis on the construction of a national history that legitimized its rise to power and produced mass consent. Organizing masses was vital because Italy lacked the economic and institutional structures that fascism deemed necessary for the establishment of a conservative civic culture. Most of all, Mussolini was aware of the need to provide Italy with a strong national narrative that rooted fascism itself in Italian history, and so urged fascist intellectuals and propagandists to delineate it. The key features of fascist national history were the revolutionary discontinuity between fascism and liberal Italy, and the genealogical continuity between fascism and *Risorgimento*. Until the 1930s, however, this narrative was not imposed on intellectuals as a binding theoretical framework but rather ignited in an apparently pluralist cultural space through patronage and by giving institutional support to initiatives aligned with it or prone to become so. The main reason behind this strategy was the recognition that violence and coercion against intellectuals would prove ineffective and even counterproductive, and so it was better to intervene at the institutional level. Thanks to patronage, job offerings, prize awarding, and public recognition fascism managed to create networks of influence that allowed to marginalize anti-fascism, take control of non-fascist cultural institutions and create new ones, all the while maintaining the appearance of libertarian governance.[8] Fascism, for example, did not suppress the somewhat anti-fascist Lyncean Academy but substantially deprived it of funding and created the *Accademia d'Italia* as its direct rival. The Italian geography of cultural institutions was reshaped, and fascism increasingly relied on propaganda to produce a sense of cultural legitimacy. In March 1925, the first national congress of fascist cultural institutions was held in Bologna and originated the National Institute of Fascist Culture and the *Manifesto degli intellettuali fascisti*, both explicitly aimed at overcoming the supposed antithesis between fascism and culture. The regime took control of the *Istituto Luce*, entrusting it with the mission of spreading fascist ideals through newsreels and documentaries; a number of institutions, after-work associations, and recreational clubs were founded; and in 1927, the EIAR (Ente per le Audizioni Grammofoniche) was created with the purpose of controlling radio broadcasting.

This approach allowed for the coexistence, at cultural level, of local and national (that is, fascist) narratives notwithstanding the progressive obliteration of local administrative autonomy carried out by the central Government.

---

8 On this, see Ruth Ben-Ghiat, *Fascist Modernities: Italy, 1922–1945* (Berkeley, Los Angeles, London: University of California Press, 2001), pp. 21–29; Claudio Fogu, *The Historic Imaginary: Politics of History in Fascist Italy* (Toronto: University of Toronto Press, 2003).

Even though ruled by appointed *Podestà* and not by directly-elected mayors, Towns continued to be cherished as *piccole patrie* ("little fatherlands") as they had been since the Communal era, but local heroes were now celebrated as fascist precursors and local cultural enterprises were presented as 'fascist in spirit'. Fascism assembled a set of values, myths, and narratives that could be easily superimposed on local ones and spread through patronage networks to establish a wide range of mutualistic interactions with local institutions and cultural circles: while the regime exploited them for propaganda, they could in turn benefit from its support to bring forth their narrative at national level. We will see in a moment how this approach worked in practice; for now, it is important to stress three important and related points: one, that fascism's rise to power caused a sudden substantial social change that dramatically affected cultural policies and planning even at local level, since Italian cultural institutions were required to reshape their narratives and aims if they wanted to stay alive; two, that exhibitions and celebrations became a fundamental component of a comprehensive cultural action implemented through patronage networks and ultimately aimed at mass indoctrination and social organization; three, that the mutualistic interaction established by the fascist regime with cultural enterprises and institutions would routinely tend to evolve into severe parasitism.

2 The National Scientific Heritage Protection Group

The speech delivered by Corsini in Bologna made a great impression on his colleagues – so great, in fact, that the Society unanimously agreed to address a petition to the Minister of Education, Giovanni Gentile. They asked for an extension of the scope of application of the Rosadi Act in order to include historical scientific collections, and for the appointment of competent historians of science to serve as consultants to museums and libraries. These requests would be formally presented to Gentile as a point of order by senator Silvio Pellerano in 1924.[9]

Meanwhile, back in Florence, Corsini got in touch with Superintendent of the State-owned Galleries of Florence and director of the *Uffizi* Gallery, Giovanni Poggi, to whom he expounded his plans. Poggi was an art historian who, during the 1910s, had managed the compilation of a general catalogue of the artistic manufacts held in Tuscany, and was also a supporter of the *Brigate toscane degli amici dei monumenti* (Tuscan Brigades of the Friends of

9 "Gruppo per la tutela del patrimonio scientifico nazionale – Comunicazioni ufficiali," *Archivio storia della scienza* 5 (1924), 87–88.

Monuments), a federation of several voluntaristic associations devoted to the protection of the chief monuments of their respective cities. The *Brigate* had begun forming in 1902 in Florence, Arezzo, Siena, and other Tuscan towns, and acted on the basis of the *actio popularis*, a legal principle guaranteed by the Italian law. They had had little communication with each other until 1908, when they merged into a single federation. The agreement was ratified during a meeting held in Florence in the presence of Giovanni Rosadi himself – promoter and first signer of the Act bearing his name – who recognized the important role of the *Brigate* in the elaboration of the Act.[10]

Corsini and Poggi thought that something similar could be done for historical scientific collections, too. So, in the first months of 1923, clearly drawing inspiration from the *Brigate*'s enterprise, they issued a call for the importance of protecting the scientific collections located in Florence. "We are a group of friends of science," they announced, rather unassumingly. "We propose to create in Florence an association of learned people who are willing to undertake in defense of the historical scientific heritage the same action that has successfully been taken by others in defense of Art."[11] Shortly thereafter, as positive feedbacks from many personalities of the city's political and cultural scence came in, Corsini and Poggi called a meeting.

On Thursday the 3th of May, 1923, the National Scientific Heritage Protection Group (*Gruppo per la tutela del patrimonio scientifico nazionale*) was created. During the first meeting, its affiliates approved a Charter that fully reflected Corsini's aspirations, for it defined the protection and promotion of historical scientific collections and the encouragement for the establishment of local groups throughout the country as the Group's main objectives.

> CHARTER
>
> I. It is constituted in Florence a *National Scientific Heritage Protection* Group tasked with the following aims:
>
> 1) Taking action for the most effective protection and retrieval of the national scientific heritage.
> 2) Promoting the cataloguing of public and private scientific collections to prevent their dispersion.
> 3) Convincing competent Authorities to bring such collections out of their present indecorous condition as fast as possible [...].

---

10 On the *Brigate* see Osanna Fantozzi Micali, "La tutela del patrimonio culturale fra istituzioni e cittadini," in *Beni culturali, ambiente, paesaggio. L'operato delle associazioni nell'azione di tutela*, eds. Osanna Fantozzi Micali and Elena Lolli (Firenze: Alinea Editrice, 2009), 5–7.

11 "Gruppo per la tutela del patrimonio scientifico nazionale. Comunicazioni ufficiali," *Archivio di storia della scienza*, 4 (1923), 198.

4) Compiling a list of all valuable objects and collections [...].
5) Contributing to make such collections easily accessible to the general public [...] by writing guides, organizing conferences, promoting group and school visits, publishing articles on magazines and newspapers.

[...]

8) Extending this action to other parts of Italy.[12]

Once the Charter was approved, the participants went on to elect the executive board. They elected Antonio Garbasso as president, Andrea Corsini as vice-president, Lino Vaccari as secretary, and Piero Bagnesi Bellincini as treasurer; Gino Bargagli Petrucci, Aldobrandino Mochi, Giovanni Poggi, and Giuseppe Stefanini were to serve as Councilors. The Museum of Ethnology and Anthropology, directed by Mochi, was chosen as the headquarter for the Group, and the *Archivio di storia della scienza* was designated as its official journal.

The latter decision was clearly an attempt to get support from Mieli, who, around the same time, was campaigning in Rome for the establishment of a National Library and Museum of History of Science. Upon receiving the news, Mieli, though grateful and very supportive, remarked nevertheless the necessity to coordinate their efforts by acting on the basis of a shared general plan. "The mere fact of this new entity being created," he wrote, referring to the Group, "forces us to find agreements among institutions devoted to the history of science in order to avoid any waste of energy caused by lack of unity."[13] At that time, besides the Group and the *Società di storia delle scienze mediche e naturali*, there existed at least two other major institutions someway related to the history of science: the National Institute for the History of Physical and Mathematical Sciences (*Istituto Nazionale per la Storia delle Scienze Fisiche e Matematiche*, presided over by Enriques), and the Italian Historic Institute of Healthcare Art (*Istituto Storico Italiano dell'Arte Sanitaria*, presided over by Mariano Borgatti). Both were based in Rome, and Mieli was trying to bring them into a single federation in the hope of creating a central directing institution. This was not, however, what Corsini had in mind. He thought of the Group as an autonomous entity that could fit in with the local community and define its plans accordingly. Unlike Mieli, he envisaged a decentered network of local museums, institutes, and libraries linked to each other through sharing

12 "Gruppo per la tutela del patrimonio scientifico nazionale. Comunicazioni ufficiali," *Archivio di storia della scienza*, 4 (1923), 198–199.

13 "Gruppo per la tutela del patrimonio scientifico nazionale. Comunicazioni ufficiali," *Archivio di storia della scienza*, 4 (1923), 200.

the same cataloguing and bibliographic standards while enjoying significant autonomy as for the rest.

The Group's earliest initiatives were directed towards protection and promotion. The Group partnered with *Pro Cultura* in organizing conferences and guided tours of the *La Specola* Museum, petitioned the University of Florence for the restoration of the *La Specola*'s anatomical waxes, and asked for a new room to exhibit Peter Leopold's chemistry cabinet.[14] In doing this, Corsini felt ensured by Gentile's recent response to Pellerano's point of order. Gentile had indeed confirmed that scientific materials of historical interest were included within the scope of the Rosadi Act, and that the Ministry itself was about to issue a Circular Letter urging museums, libraries, and Universities to report about the nature and condition of any historical scientific collection they held.[15] In his response, Gentile had also highlighted how difficult it was to discern the historical value of scientific collections, implying that, if the Ministry had not undertaken any particular involvement in the matter so far, it was mostly because there was no trained personnel who could effectively perform the job. To Corsini, this sounded like a confirmation that the Group should broaden its purposes beyond safeguarding and promoting to include research and consultancy. According to him, the Group could assist local institutions in locating collections, inspecting them, and eventually preparing detailed reports to be sent to the Ministry.

Corsini had already presented an ambitious proposal during the meeting of the Group held on November 28, 1923. His idea was to get local museums involved by asking them for catalogues of their collections – or, at least, for detailed descriptions of them – to be perused by the Group in search of items of historical scientific interest. To serve this goal, Corsini noted, the Group itself needed a research library, however small, to gather books, pictures, drawings, prints, objects – in brief, anything whatsoever related to historical scientific collections.[16] It was a significant step forward compared to the *Brigate* experience: in fact, the Group presented itself not as a voluntary organization mostly intent upon sponsoring restoration projects and celebratory initiatives, but rather as a partnership of professionals whose mission was to protect and promote the Italian scientific heritage through historical investigations.

---

14 "Gruppo per la tutela del patrimonio scientifico nazionale. Comunicazioni ufficiali," *Archivio di storia della scienza*, 6 (1925), 96.

15 "Gruppo per la tutela del patrimonio scientifico nazionale. Comunicazioni ufficiali," *Archivio di storia della scienza*, 5 (1924), 316.

16 "Gruppo per la tutela del patrimonio scientifico nazionale. Comunicazioni ufficiali," *Archivio di storia della scienza*, 5 (1924), 87.

However, the most ambitious proposal was launched by Garbasso during the very first meeting of the Group, in 1923: an exhibition of the historical scientific collections held in Tuscany, to be staged from March 15 to May 15, 1924. Such an event, according to Garbasso, would draw attention to the Group's activities and strongly substantiate their claims. Speaking in his capacity as mayor, he assured that the City of Florence would put Palazzo Pitti at their disposal.[17]

3 The Galilean Tradition and the Florentine Primacy

In 1923, Garbasso was 51 years old and had been serving as the mayor of Florence for three years (he would keep that position, albeit as *Podestà*, until 1927). He was a famed physicist – actually, one of the greatest of his generation. After graduating from the University of Turin, he moved to Bonn to study with Heinrich Hertz, and then to Berlin, under the guidance of Hermann von Helmholtz. Back in Italy, he taught Experimental Physics in Turin, Pisa, Genoa, and Florence. A veteran of the Great War, he joined the fascist party in 1923, and the following year was nominated to Senate. In the aftermath of Matteotti's murder, Garbasso played a key role in bringing the Florentine section of the Italian Liberal Party to Mussolini's side, thus gaining the *Duce*'s personal gratitude and a plethora of public offices related to the administration of scientific research. He presided over the Italian Society of Physics and the *Accademia d'Italia*, and promoted the creation, in Florence, of the Optics Laboratory (1918) and of the Institute of Physics (1921). Furthermore, he was the director of the *Museo degli strumenti antichi*.[18]

Garbasso considered history of science mostly as an instrument of social education and as a useful tool of political and ideological propaganda. He was convinced that a strong attitude towards 'realism' was inherent in the Italian character, and, therefore, that Italians themselves were 'naturally inclined' to engage in experimental science. Galileo Galilei, he maintained, stood out as the prototype of the 'Italian genius' precisely because of his strong sense of reality in practicing science, his mathematical rigor in philosophical reasoning, and his devotion to the Church in submitting himself to its decrees. If these were the qualities defining the 'Italian genius', then, for Garbasso, the 'Galilean tradition' basically summed up the whole Italian scientific tradition. Therefore, in

17 "Gruppo per la tutela del patrimonio scientifico nazionale. Comunicazioni ufficiali," *Archivio di storia della scienza*, 4 (1923), 200.

18 For a biographical survey, see *DBI*, "Garbasso, Antonio," by Giulio Peruzzi.

addition to being the cradle of the Italian language, Florence could also rightly lay claim to be the cradle of Italian science. "Throughout the 17th century," he argued, "physics was an Italian, or rather a Florentine, science. Galileo founded Mechanics and Astrophysics; Torricelli, Terrestrial Physics; the Cimento Academicians, Experimental Physics."[19] Every great advance in science, according to him, could be traced back to the 'experimental method' envisaged by Florentine physicians like Taddeo Alderotti in the Middle Ages, then enhanced by Leonardo during the Renaissance, and eventually developed in full by Galileo in the Seventeenth Century. On the basis of these ideas, he campaigned hard, especially after 1923, for the popularization of science, stressing, among other things, the exemplary value of history and exhibitions. Awakening people's love for science was, for him, the same as awakening their love for their homeland.[20]

When Corsini and Poggi reached out to him, Garbasso had been managing the *Museo degli strumenti antichi* for almost ten years and had great projects in mind. He dreamt of a great "Galilean Museum" based in Arcetri, which, if realized, would have comprised the Optics Laboratory, the Institute of Physics, the Astronomical Observatory, and a brand-new historical museum to be established at the villa *il Gioiello* (the Jewel – Galileo's deathplace). The villa was actually a private property, but in 1921 the Italian Government, following Garbasso's advice, had declared it a National Monument, putting it under the tutelage provided by the Rosadi Act.[21] That meant that the Italian State had a pre-emption right, and Garbasso proposed it to be exercised as soon as possible. To him, the villa was the perfect place to house the *Museo degli strumenti antichi* and give life to a true "Science City" wherein the past and present of the Italian scientific tradition were symbolically reunited under the sign of Galileo.

Not everyone, in Florence, agreed with Garbasso. Physicist Carlo Del Lungo, for example, thought that *il Gioiello* was too far away from Florence urban area

19 Antonio Garbasso, "Il contributo degli Italiani alla fisica," in *L'Italia e la scienza*, edited by Gino Bargagli Petrucci (Firenze: Le Monnier, 1932), 67–96: 84.

20 Garbasso expounded his ideas on this subject in a series of pamphlets and lectures mostly published between 1910 and 1920, such as "Parole del prof. Garbasso," in *Atti della società italiana per il progresso delle scienze, VI riunione* [excerpt]; "La tradizione del pensiero toscano: discorso inaugurale letto nella sala di Luca Giordano nel palazzo mediceo Riccardi il 6 novembre 1916 dal prof. Antonio Garbasso," in *Annuario del R. Istituto di Studi Superiori pratici e di perfezionamento* [excerpt]; "L'orma di Galileo: una visita ai laboratori di Arcetri," *Nuovo Patto*, 5 (1922); "La scienza e la filosofia nostra nazionale. Discorso letto all'adunanza solenne del 4 giugno 1922 dal socio Antonio Garbasso," *Atti dell'Accademia Nazionale dei Lincei*, 3, VIII [excerpt].

21 Antonio Favaro, "La casa di Arcetri dove spirò Galileo «Monumento Nazionale»," *Archivio di storia della scienza*, 2 (1920), 438–440.

to successfully attract visitors, and insisted on the fact that Palazzo Torrigiani was already an "appropriate and beautiful seat" instead. Garbasso replied, affirming that in Palazzo Torrigiani there wasn't enough room to set up a historical museum like the one he had in mind, that is, an institution more or less like the *Deutsches Museum* – a place where visitors could admire the progressive development of scientific instrumentation and possibly see machines and apparatuses in action.[22]

Either way, Garbasso would not fulfill his dream. The Government was unwilling to buy the villa and Garbasso would soon realize that the City of Florence alone could not afford to fund his project. He himself, during the 1920s, first as mayor and then as *podestà*, would cut funding for social services, public instruction, and council housing in order to restore budget balance. Moreover, early Florentine fascism was dominated by the riotous and restless violence of a *squadrismo* that looked suspiciously at personalities like Garbasso – a catholic, moderate intellectual who had reportedly joined fascism only after the *Squadre* themselves had completed 'the capture of Florence'. In 1927, the weekly satirical magazine *il Brivido* featured on its front cover a caricature of Garbasso entering Florence in Renaissance costume, while the caption labelled him as the "Turin braggart."[23] According to historian Marco Palla, as *podestà* Garbasso was "a purely decorative figure."[24] Nevertheless, in 1928 he would try to resume his plans. That year, he would ask the University of Florence for the concession of the historical collections held in the *Museo degli strumenti antichi* with the purpose of establishing a museum. This time, Garbasso was planning to set it up not at *il Gioiello*, but at the *Giardino dell'Orticultura* (Horticulture Garden), a park owned by the Tuscan Society of Horticulture, which was negotiating with the City of Florence for its sale.[25] But lack of funding, again, would prevent the City Council from sustaining him, and so Garbasso would definitively give up on his project and decide to fully support Corsini's one.

22 On this debate, see Carlo Del Lungo, "Il R. Museo di Fisica e storia naturale di Firenze: quello che è e quello che dovrebbe essere," *il Marzocco*, 27, n. 33, August 13, 1922; Antonio Garbasso, "Per il R. Museo di Fisica e Storia Naturale di Firenze," *il Marzocco*, 27, n. 35, August 20, 1922; Carlo Del Lungo, "Ancora a proposito del Museo di Fisica e Storia Naturale di Firenze," *il Marzocco*, 27, n. 35, August 27, 1922.

23 "L'apoteosi," *Il Brivido*, III, 2, 1927.

24 See Marco Palla, *Firenze nel regime fascista. 1929–1934* (Firenze: Leo S. Olschki, 1978), 94–95; Roberto Cantagalli, *Storia del fascismo fiorentino 1919/25* (Firenze: Vallecchi, 1972), 316.

25 Garbasso to Burci, May 18, 1928, ASUFi, a. 1930, f. 597, fasc. 126.

However, it is important to note here that, before Garbasso got involved, neither the identification of the Italian scientific tradition with the Galilean one nor the claim of a Florentine primacy had had any place in Corsini's project. Though acknowledging Galileo's leading role in the development of experimental science, Corsini had never went so far as to identify the Galilean tradition with the Italian tradition. On the contrary, he had often remarked the importance of giving "secondary" figures the credit they deserve. "In science, either you are a Galileo or you are nothing," he had lamented in 1922 at the Bologna meeting of the *Società di storia critica delle scienze mediche e naturali*. For him, Galileo was undoubtedly one of the greatest minds in the history and immensely contributed to the development of science in Italy, but this did not mean that Italian science was, in itself, "Galilean in its essence." Still, once Garbasso entered the scene, ideological elements of this kind were increasingly given more publicity, until, after the involvement of Piero Ginori Conti, they came to the foreground.

4 The Institute of the History of Science in Florence

Corsini's project of an index of the historical scientific collections held in Tuscany soon proved more difficult to realize than expected. First of all, a suitable operational headquarter was needed, but facilities at the Museum of Ethnology and Anthropology were limited to a shared meeting room. Moreover, the Group's internal regulations provided that, in the event of the *Gruppo* being dissolved, all its reference material was to pass to some institution devoted to the history of science. In fact, since no such institution existed in Florence, that meant that all the work done by the Group would have very likely been lost.[26] So, scarcely a year after the project had been launched, Corsini turned to the University of Florence for support. It was not, at this first stage, a matter of funding: what he actually wanted was an institutional basis.

Towards the end of 1924 or the beginning of 1925, Corsini addressed a petition to the rector of the University of Florence, the physician Giulio Chiarugi, expounding his proposal for establishing an 'institute of history of science' within the University. Corsini made it clear that there was no need for money, since a medium-size, passably furnished room, together with the services of a keen librarian, would suffice for his purposes. The rector answered on March 7, 1925, announcing that, after a consultation, the Academic Senate had

26 "Gruppo per la tutela del patrimonio scientifico nazionale. Comunicazioni ufficiali," *Archivio di storia della scienza*, 4 (1923), 200.

agreed to grant Corsini a room in via degli Alfani to set up a small library with the assistance of a University Librarian, Ms. Rosa Franceschini.[27] As soon as he received Chiarugi's communication, Corsini had everything pertaining to the Group transferred to the new premise. Then, together with Ms. Franceschini, he began to compile an inventory.

The Institute of the History of Science was officially inaugurated on May 7, 1925, in the presence of Chiarugi, and five months later Corsini sent to the rector his first report on its activities. At that time, the Institute held around 120 printed books, 100 manuscripts, and 1000 booklets. Corsini and Franceschini had collected almost 500 photographs and had been working on a bibliography which counted over 2900 entries so far. Moreover, they were busy negotiating with the Santa Maria Nuova General Hospital to entrust the Institute with the conservation of the Hospital's historical surgical instrument collection. The Institute, according to Corsini, were performing very well and was earning reputation, even though it had no funds and consisted of but one single room. Skillfully taking advantage from the University's aspiration to excellence and renown,[28] Corsini concluded his report by stating that the Institute could easily become the "best institute and museum of the whole world" had the University itself wanted it to. There were plenty of scientific instruments no longer in use stashed in the University's warehouses to set up a beautiful museum, Corsini noted, and what they needed was just a strategic plan to get support from public and private institutions.[29] On December 4, 1925, Chiarugi wrote back. The rector offered Corsini his congratulations, and let him know that the idea of a strategic plan for the further development of the Institute in the interest of the University of Florence had been discussed in an Academic Senate hearing. The University, Chiarugi went on, would appoint an Evaluation Commission, consisting of Giorgio Abetti, Giovanni Poggi, and Piero Ginori Conti, tasked with advising on the most effective way to proceed.[30]

If Abetti (director of the Arcetri Astronomical Observatory) and Poggi were longtime acquaintances of Corsini and had been dealing, in some way or another, with the matter in hand, Ginori Conti was an outsider, apparently. Still, he was chosen on purpose. A wealthy nobleman of ancient lineage and

27 Chiarugi to Corsini, March 7, 1925, ASUFi, a. 1925, f. 556, fasc. 126.

28 The University of Florence had been established in 1924, when the *Istituto Superiore di Studi Pratici e di Perfezionamento*, founded in 1859, was officially granted the status of University. The very first *Studium generale* had been founded in 1321, but lasted intermittently until 1515, when the Medici elected the University of Pisa as the Tuscan Granduchy University.

29 Corsini to Chiarugi, October 29, 1925, ASUFi, a. 1925, f. 556, fasc. 126.

30 Chiarugi to Corsini, December 4, 1925, ASUFi, a. 1925, f. 556, fasc. 126.

refined taste, well-read and interested in science, Piero Ginori Conti, Prince of Trevignano, was a competent bibliophile, a generous patron of arts, and a charitable philanthropist. He was also very fond of history, and was himself an amateur historian, having written on Savonarola, on the Medici Family, and on Pietro Alighieri. Born in Florence in 1865, after graduating from the School of Social Sciences in Florence he had married the Countess Adriana de Larderel, heir to the *Società Boracifera di Larderello*, a boric acid production company based in Val di Cecina. Once appointed General Manager of the *Boracifera* by his father-in-law Florestano de Larderel in 1904, he undertook a series of systemic reforms in order to modernize and expand the company. He established an in-house research laboratory of chemistry to strengthen ties with scientific research and appointed Raffaello Nasini, a famed chemist from the University of Pisa, to manage it. The company's area of operations was enlarged to include production of electricity from geothermal energy, extraction of helium from natural gas, and the manufacturing of ammonia. Within a few years, the *Boracifera* became a roaring international success and Ginori Conti one of Italy's most recognized businessmen.[31]

The Prince could wield strong political influence, too, for he had entered politics in 1896 and in 1900 had been elected to the Parliament. In 1919, he had enthusiastically adhered to fascism, which he saw as the force of order and restoration that could ensure social peace and, consequently, entrepreneurship development. Between 1919 and 1920 he had promoted the creation of several fascist squads – even allowing his own son, Giovanni, to join up with them – and had the *Boracifera* workers participate as Blackshirts in the *Marcia su Roma*. During the years of fascist rule, he served as senator and got friendly with Mussolini, to whom he used to pay visit each time he happened to be in Rome. Given such circumstances, the Academic Senate's choice looks quite felicitous. Ginori Conti was learned, shrewd, powerful, wealthy, generous – in brief, anything one could ask for from a patron. He and Corsini got along well, and eventually a close friendship based on mutual esteem grew up between them. The Prince let himself became infected with Corsini's enthusiasm, never failing to give political and financial support to the physician's initiatives, even though Corsini did not share the Prince's political view.

A few months after taking office, the Evaluation Commission presided by Ginori Conti sent its report to the new rector, Enrico Burci. The Commission largely agreed with Corsini's opinion: if supplied with adequate funding, the

---

31 On Piero Ginori Conti, see the "Ginori Conti, Piero" entry in *DBI* (by Fulvio Conti), and Michele Lungonelli, Martina Migliorini, *Piero Ginori Conti. Scienza, cultura e innovazione nella Toscana del Novecento* (Roma-Bari: Laterza, 2003).

Institute did have substantial room for development. Ginori Conti had donated 10.000£ out of his own personal funds, but the Institute had to seek for public funding. So, according to the Commission, the first step to be taken was to ask for public patronage by applying for the recognition of the status of *Ente Morale* (that is, juridical personhood). As for the premise, the Commission suggested to relocate the Institute at the *Tempio degli Scolari*, a small building designed by Filippo Brunelleschi and located in the neighborhood of Via degli Alfani, a urban area that the City of Florence was planning to renovate. Finally, the commissioners communicated that, in order to gain support from privates, they had decided to revive the Group's project of an exhibition of historical scientific instruments and were currently negotiating with the Government to this end.[32]

Burci, approving all of this, got the application process started. Ginori Conti, Abetti, and Poggi drafted a preliminary charter, while Corsini drew up a report about the Institute's mission, activities, and goals. The Institute, Corsini wrote in his report, sought to arrange a bibliography of history of science, to collect an iconography of scientists, to gather scientific instruments and machines, and to set up a history of science library. "Over the last year," he wrote, "the Institute, though lacking any funding whatsoever, did its best to accomplish these ends. It has already borrowed from privates a number of objects, manuscripts, paintings, books, etc., and has drafted an index [...] of works on history of science published in Italy and abroad." Corsini was careful enough to signal the eminently patriotic purpose of the Institute, too:

> Until now, Italy lacked an Institute of History of Science. The R. University of Florence has been the first one to establish it, in 1925, keeping up with a scientific movement that is getting stronger everywhere and is of primary importance for Italy, because Italy itself is the country to which is owed that flourishment of Science experienced in the Renaissance and in the XVIII century, and so Italy itself can very justly be considered as the cradle of modern science. Apart from the importance in itself of the history of science, valuing what in science has been done by Italians is nowadays necessary against the foreign publications which neglect and overlook the Italian genius' feats [...].[33]

It is worth noting that this kind of 'scientific nationalism' was absent both from Corsini's and the Group's original program. It is clear that Garbasso's and Ginori Conti's enthusiastic adhesion to fascism had had consequences, but it is also

32 Ginori Conti to Burci, n.d., February, 1926, ASUFi, a. 1926, f. 564, fasc. 126.

33 Corsini to Burci, n.d., April 1926, ASUFi, a. 1926, f. 564, fasc. 126.

clear that, since 1923, things had profoundly changed. Although nationalism – and scientific nationalism along it– had begun to grow well before Mussolini came to power, fascism gave momentum to and profiteered from it, especially after 1925. As the regime tightened its grip on Italy, 'scientific nationalism' was given more and more room. Fascism fostered it for propagandistic purposes, while Corsini was willing to endorse it because he was convinced that in this way the regime would back him up with his projects. Corsini's partnership with Garbasso and Ginori Conti, thus, assured him institutional support, but, on the other side, as we shall see, brought a strong ideological and political commitment with it.

The Institute of History of Science was recognized as *Ente Morale* in 1927. Its board consisted of personalities from the University of Florence (Giotto Dainelli, a geographer and geologist, and the chemist Guido Pellizzari), the Ministry (Senator Girolamo Gatti), the Private Donors (Piero Ginori Conti), and the City of Florence (Antonio Garbasso). Corsini was elected as director.[34] Being an entity administered by and directly linked to the University, the Institute was different both from the Institute for the History of Physical and Mathematical Sciences of Rome, which offered undergraduate courses only, and from the Historic Institute of Healthcare Art, which was mainly a museum: the Florentine institution focused on conservation and scholarly research, in the wake of what Karl Sudhoff and Charles Singer – with whom Corsini had corresponded during the '10 – had done in Leipzig and Oxford, respectively.[35] With the official birth of the Institute, the Group gradually slackened its activities, and eventually dissolved after 1929.

5 Aldo Mieli and the National Institute of History of Science in Rome

One "single room" wherein "the director's enlightened work creates the Institute's various branches of activity and materializes them in collections, books, illustrations, catalogues, etc."[36] It was with an ill-concealed irony that Mieli, in 1925, informed the readers of the *Archivio* about the creation of

34 Burci to Corsini, February 4, 1928, ASUFi, a. 1927, f. 571, fasc. 126.

35 Corsini had also visited Klebs in Leipzig. See Andrea Corsini, "L'Istituto di Storia della Medicina della Università di Lipsia," *Rivista di storia delle scienze mediche e naturali*, 1919, 165–166. On the whole matter, see Marco Beretta, "Andrea Corsini and the Creation of the Museum of the History of Science in Florence (1930–1961)," in *Scientific Instruments on Display*, ed. by Silke Ackermann, Richard L. Kremer, Mara Miniati (Leiden-Boston: Brill, 2014), 1–36: 5–6.

36 Aldo Mieli, "L'Istituto di storia delle scienze nell'Università di Firenze," *Archivio di storia della scienza*, 6 (1925), 198–199.

the Institute of History of Science in Florence. He could hardly be blamed for his irony: the Institute at that time truly consisted of but a single room which was too warm in the Summer and too cold in the Winter.[37] Still in 1927, Corsini himself would acknowledge that the Institute, despite having such an "high-sounding name," was "a very little thing, actually."[38] It had arisen quietly, thanks to the willpower of a bunch of learned men whose aim was to protect historical scientific collections, and, until Ginori Conti got involved, it was supposed to be just a facility to serve the Group.

On the other hand, during those very same years Mieli was busy trying to realize in Rome a National Library and Museum of History of Science following a radically different approach. For him, the first step to be taken was to reunite in a federation the Italian associations and institutions devoted to the history of science.[39] In Mieli's plan, this federation should coordinate the members' initiatives acting as their institutional representative in the dealings with political institutions. In 1924, he called a meeting of representatives of the National Institute for the History of Physical and Mathematical Sciences, the Italian Historic Institute of Healthcare Art, the National Scientific Heritage Protection Group, the History of Medical and Natural Sciences Society, the Italian Geographical Society, and several other institutions and individuals variously interested in history of science.[40] During the meeting held in Naples on May 2, 1924, they agreed to constitute the National Federation of Societies, Institutions, Teachers, and Lovers of History of Science (*Federazione Nazionale fra le Società, gli Enti, gli Insegnanti ed i Cultori di Storia della Scienza*). Its elected board, which featured representatives from all the adhering societies,[41] officially met for the first time on November 22 in Rome. On that occasion,

---

37 Corsini to Burci, February 23, 1926, ASUFi, a. 1926, f. 564, fasc. 126.

38 Andrea Corsini, "Firenze e la storia delle scienze," *Atti della società colombaria di Firenze*, February 20, 1927, 256–263.

39 Aldo Mieli, "Per un'intesa fra le società di storia della scienza e l'Archivio," *Archivio di storia della scienza*, 4 (1923), 109–110.

40 Aldo Mieli, "Per un'intesa fra gli studiosi e gli enti di storia della scienza," *Archivio di storia della scienza*, 5 (1924), 79–81.

41 Specifically, the board consisted of Roberto Almagià (Italian Geographical Society), Silvestro Baglioni, Guglielmo Bilancioni, and Davide Giordano (History of Medical and Physical Sciences Society), Giovanni Battista De Toni† (Royal Commission on Vincian Studies), Ugo Vram (Italian Anthropological Society), Mariano Borgatti (Italian Historic Institute of Healthcare Art), Nicola Vacchelli (Military Geographic Institute), Gino Loria (editor of the *Bullettino di bibliografia e storia delle scienze matematiche*), Andrea Corsini, Antonio Garbasso, Federigo Enriques, and Aldo Mieli. See "Federazione Nazionale fra le società, gli enti, gli insegnanti ed i cultori di storia della scienza. Comunicazioni ufficiali," *Archivio di storia della scienza*, 5 (1924), 189–193.

Mieli, after having assured once again that the Federation would not interfere with the adhering societies' internal administration, pointed out its most immediate aims:

> As for now, what we need to start out is just a large enough place to settle in, an attendant, and a cleaner. The Societies could take their book collections there, each in its own assigned space, so that the said collections could be easily consulted [...]. We could gather in that place old instruments and other stuff which now burden scientific institutes as a worthless encumbrance; we could also house there the remains of the beautiful and useful Copernican Museum. [...] Private collections would be welcome, too.[42]

The establishment of a 'National' Library and Museum of history of science was, for Mieli, the most urgent need, and therefore the Federation should devote all its efforts to it. Unfortunately for Mieli, in the first two years of its life the Federation did not work very well. 'Federated' societies went on with their usual activities without caring much about coordinating each other – in fact, they were quite unwilling to undertake any common action. Enriques had never being present at a Federation meeting in two years, Corsini had founded the Institute of History of Science, Garbasso fancied the creation of a 'Science City' near Florence, and Pietro Capparoni and Mariano Borgatti were studying strategies to create a National Museum of the History of Medicine. And all of this was happening without the Federation being involved in any respect. It looked like the Federation was nothing but "another society" actually aggravating that "energy wasting" often lamented by Mieli.[43]

Fascism's new drive towards centralization, Mieli thought, offered opportunities for those who were able to grasp them. The Federation meeting held on February 1, 1926, featured, much to the attendants' surprise, Gentile and Enriques as keynote speakers. Before resigning as Public Education Minister in 1924, Gentile had carried out a comprehensive school reform which, among other things, introduced some formal teachings of history of science in high schools. Gentile's reform had been implemented in 1923, and that same year

42 "Federazione Nazionale fra le società, gli enti, gli insegnanti ed i cultori di storia della scienza. Comunicazioni ufficiali," *Archivio di storia della scienza*, 5 (1924), 397–403, 400.

43 During the Federation meeting held on February 1, 1926, Mieli deplored the fact that "privates" often decided to take initiatives such as commemorations of scientists and like without informing the Federation first. See "Federazione Nazionale fra le società, gli enti, gli insegnanti ed i cultori di storia della scienza," *Archivio di storia della scienza*, 6, (1925), 397–398, 398.

Enriques had founded the National Institute for the History of Physical and Mathematical Sciences. A year later, the two had collaborated in creating a graduate school at the University of Rome to train future history of science teachers. In order to get institutional recognition and funding, this graduate school should present itself as a 'national' entity officially involving the majority of the Italian societies and institutions devoted to history of science, and so now there they were, trying to persuade the Federation to join them in this. During his speech, Gentile warned against isolated initiatives and stressed on unity and cohesion, arguing that the Government was willing to consider supporting only a national, highly centralized institution exerting strong decision-making power, something similar to the newly established *Consiglio Nazionale delle Ricerche* (National Research Council – CNR).[44] Creating such an entity was, according to him, the only way they could get a National Library and Museum of History of Science. The Federation decided to nominate a Committee to evaluate Gentile's proposal.[45] At the end of a series of 'informal' and 'friendly' gatherings, the Committee approved the Federation's merger with the National Institute for the History of Physical and Mathematical Sciences to create the National Institute of History of Science (*Istituto Nazionale di Storia della Scienza*). The news was broken during the Federation meeting held in Palermo on May 26, 1926.[46]

Mieli was delighted. With its solid institutional recognition, the National Institute could undertake the "wide-range researches and studies" he had been longing for since the '10; it could also effectively coordinate the activities of local institutes. In a harsh editorial published in 1926 on the *Archivio di storia della scienza* and significantly titled *Fronte unico* ("Unique frontline"), Mieli stigmatized the behaviour of the "handful of people" who still "allow themselves to do a hundred things independently of one another." They, Mieli wrote, "still want *their* journal and *their* collections, and they still want privates and the Government itself to fund *their* institution."[47] Well, those times, he proclaimed, were over. The Institute was intended to provide official voice, unique

44 The CNR had been founded in 1923 to coordinate and direct Italian scientific research, and to serve as consultant to the Italian Government.

45 "Federazione Nazionale fra le società, gli enti, gli insegnanti ed i cultori di storia della scienza. Comunicazioni ufficiali," *Archivio di storia della scienza*, 6 (1925), 397–398. The Commission consisted of Enriques, Mieli, Giordano, Baglioni, and Capparoni.

46 "Istitituto Nazionale di Storia della Scienza. Comunicazioni ufficiali," *Archivio di storia della scienza*, 7 (1926), 183–184.

47 Aldo Mieli, "Fronte Unico," *Archivio di storia della scienza*, 7 (1926), 331–333. Emphasis is in the original. The title "Unique Frontline" evoked a passage from a speech delivered by Mussolini during the opening ceremony of the Italian Institute of Exports, inaugurated that same year.

direction, and systematic organization to Italian history of science as a whole, pretty much like National Institute of Archaeology and History of Art was a recognized graduate school that awarded degrees and fellowships and served as the Government's main consultant in anything relating to art and archaeology.[48] Unfortunately, such an enthusiasm prevented Mieli from seeing the imbalance of power in the Institute's administration. Indeed, Enriques acted both as Director of the Scientific Committee and as Chairman of the Board of Directors, besides maintaining his office as director of the graduate school in history of science. A few years later, from his exile in Paris, Mieli would recall all this as a big misjudgment on his part. Writing to Corsini, he would regret his choices and acknowledge as a matter of fact that, in 1926, by endorsing the merger he had ingenuously delivered the Federation into Enriques's – and consequently the regime's – hands.[49]

Corsini never endorsed either Mieli's or Enriques's project. Though formally a member of the Federation Board and, later, of the National Institute's Scientific Committee, he never attended any of the Federation meetings. He was kept informed about what was going on by his friend Davide Giordano, who, right after the Federation's merger with the Institute for the History of Physical and Mathematical Sciences, sent him a letter inviting him to adapt himself to the circumstances and accept the seat on the Scientific Committee they were offering him.[50] Though politely accepting the offer, Corsini could not resign himself to the situation. What he feared the most was that the supposedly powerful National Institute would suffocate the weak Florentine one. After all, Mieli had made it clear that, following the birth of the National Institute, peripheral centers should devote themselves to "regional and local research" only. Furthermore, Corsini felt threatened by the prospect of setting up a National Museum of History of Science in Rome, because that could result in being forced to hand Florentine collections over to it.

In response to all this, the strategy deployed by Corsini, Ginori Conti, and Garbasso was going to be lucid and effective. Building on the distinction between local and national research, relying on Garbasso's ideas about the "Galilean" tradition, and exploiting the fascist regime's need for cultural

48 "Memoriale," *Archeion*, 8 (1927), 232–235.

49 See Giovanna Baroncelli, Massimo Bucciantini, "Per una storia delle istituzioni storico-scientifiche in Italia: l'Istituto e Museo di storia della scienza di Firenze," *Nuncius*, 5, 11 (1990), 5–52, 27–29. Mieli's bitter regret notwithstanding, it would be unfair to dismiss Enriques' project merely as a 'scheme'. Enriques considered history of a science as the discipline that could link the 'humanistic' and 'scientific' cultures, and so he hoped sincerely to establish an actual research center and museum.

50 Baroncelli, Bucciantini, "Per una storia delle istituzioni storico-scientifiche in Italia," 28.

recognition they would claim and emphasize the identity of the Italian scientific tradition with the "Galilean" one, so that Florentine "local" tradition would end up being the "general" Italian tradition. On this basis, they would argue that an actual Museum and Institute of History of Science could be established in Florence only. Furthermore, they would prove all this not by means of writings and debates, but by setting up a massively spectacular event; an event able to catch the attention of the "new" *fascistissimo* regime: a majestic national exhibition of history of science.

CHAPTER 3

Setting Up the Exhibition

1 From a Tuscan Exposition to the National Exhibition

As seen from Florence, the events that led to the creation of the National Institute of History of Science had all the appearances of a gathering storm threatening to overtake the helpless Florentine Institute. Corsini, Ginori Conti, and Garbasso knew that the National Institute was a decisive step towards Mieli's centralistic project, a project within which, they supposed, their Institute featured, at best, as a modest bibliographic office. Gentile had made it clear: creating a central, empowered institution was the only effective way to get support from the fascist regime. It was apparent that the Florentine Institute of History of Science could not aspire to such a status, resting as it was on so unstable ground. In 1926 the Institute was still waiting to be recognized as *Ente Morale* and survived solely thanks to Ginori Conti's charitable contributions. For that matter, even more menacing to the Florentines – and especially to Garbasso, director of the *Museo degli strumenti antichi* – sounded the project, announced during the Federation's first meeting, of setting up a National Library and Museum of History of Science in Rome, as this could result in being forced to hand Florentine collections over to the new institution. What had happened to the *Museo Telegrafico* – established in Florence in 1878 by Ernesto D'Amico and then, in 1907, transferred to Rome – showed that such a possibility was far from being remote.[1]

To forestall the 'Roman threat', Corsini, Ginori Conti, and Garbasso agreed upon the necessity of appealing directly to the public and, by means of it, to the Government. Relying on the regime's willingness to support any initiative that could be exploited for propagandistic purposes, they decided to revive the Group's (and Institute's) project of a Tuscan exposition of historical scientific collections, enlarging it to become a proper national exhibition of history of science. In fact, there were evidences suggesting that this might be an effective strategy, as several museums borne in the wake of the success of temporary exhibitions existed in Italy. The National Museum of Popular Traditions,

1 See Leonardo Lucci, Antonio Salvini, Massimo Temporelli, Barbara Valotti, "Musei, collezioni e fonti documentali per la storia delle telecomunicazioni in Italia," in *Storia delle telecomunicazioni*, ed. by Gabriele Falciasecca, Virginio Cantoni, Giuseppe Pelosi (Firenze: Firenze University Press, 2011), 939–963: 943–944.

© KONINKLIJKE BRILL NV, LEIDEN, 2022 | DOI:10.1163/9789004297418_004

for example, had been established in Rome after the Ethnographic Exhibition of 1911; the Colonial Museum of Rome was set up in 1923 gathering the collections that had been on display at the Maritime Exhibition of Genova in 1914; and in 1925, the National Didactic Exhibition had laid the groundwork for the foundation of the National Didactic Museum of Florence.[2] The idea was to set up an event that could tickle the public's collective imagination, spectacularly substantiating the claim that Florence was the most suitable city to host a National Museum of History of Science. Florence had been "the very first city" in Italy to host a museum of history of science and to launch a protection policy for historical scientific collections, Corsini remarked while presenting the project at *La Colombaria* Academy in Florence, "therefore," he argued, "due to its history and its collections, it is clear that only Florence, in Italy, deserves to become the main center of these studies."[3]

These proposals found fertile ground in the Florentine cultural and political environment. A small city of about 300.000 inhabitants, Florence was trying to redefine its identity as the Italian Capital of Art and Culture. Lacking an industrial economy on the order of northern cities like Milan and Turin, and still facing financial difficulties after the expenses incurred during the brief period it had served as national capital of Italy (1865–1871), the city increasingly relied on tourism as a means of economic development. Among the many initiatives flourished during the 1910s and the 1920s, the most significant had surely been the foundation, promoted in 1923 by journalist Enrico Barfucci, of the *Ente per le Attività Toscane* (EAT-Authority for Tuscan Activities). Though officially independent and non-political, the EAT was in fact a partly state-controlled institution backed by the fascist party of Florence and its patronage network. Its board featured intellectuals like the aforementioned Giovanni Poggi, the writer Giovanni Papini, and the publisher Enrico Bemporad as well as politicians like Alessandro Martelli – the future Minister of National Economy – Italo Capanni, deputy and founder of the first fascist formation of Florence, and Carlo Delcroix.[4] Alessandro Pavolini, the rising star of Florentine fascism who would become Provincial Federal Secretary of the Fascist Party in 1929, though not directly involved in the EAT's management, was also amongst its

2 On these initiatives, see Sandra Puccini, *L'itala gente dalle molte vite. Lamberto Loria e la Mostra di Etnografia italiana del 1911* (Roma: Meltemi editore, 2005); Francesca Gandolfo, *Il Museo Coloniale di Roma (1904–1971). Fra le zebre nel paese dell'olio di ricino* (Roma: Gangemi editore, 2014); *Radici di futuro: l'innovazione a scuola attraverso i 90 anni dell'Indire. Catalogo della mostra*, ed. by Pamela Giorgi (Firenze: Indire, 2015).

3 Corsini, *Firenze e la storia delle scienze*, cit., 8–9.

4 On the EAT see Anna Pellegrino, *La città più artigiana d'Italia. Firenze 1861–1929* (Milano: FrancoAngeli, 2012), 80–82.

supporters. In an article published on *Battaglie Fasciste* ("Fascist Battles") in 1926 and titled *Per la più grande Firenze* ("For a bigger Florence"), he pointed out the importance of cultural tourism for the city's local economy, insisting on the necessity of "fighting" other Italian cities in a friendly competitive spirit to show Florence's cultural primacy, thus certifying its status as "Capital city of the Italian brightness." In a sense, Pavolini's political action during the 1920s and 1930s embodied fascism's whole attitude toward local traditions and cultural enterprises. In 1929, he created *il Bargello*, a weekly magazine that soon became a focus in articulating local and fascist narratives in order to establish mutualistic interactions between the regime and preexisting local cultural enterprises. Thanks to Pavolini's efforts, initiatives such as the International Book Fair or the Art Exhibition *la Fiorentina primaverile* were revived and celebrated as fascist events testifying the glory of the new 'fascist' Florence.[5]

In 1926, the EAT was busy organizing an historical exhibition of illustrated books, so it is no surprise that Barfucci and his associates got interested in the project of a national exhibition of history of science and eagerly agreed to give their support. For the EAT, establishing a National Museum of History of Science alongside the National Library and the Uffizi Gallery would have meant restoring the "Tuscany's intellectual exarchate" eulogized by Vincenzo Gioberti eighty years earlier.[6] Thus, on February the 26th, 1927, Corsini presented the initiative at the EAT headquarters in via de' Ginori, stressing on "the role played by Italy and, specifically, Tuscany," in history of science. After a brief consultation, the "Executive Committee of the First National Exhibition of History of Science" was created. It featured Ginori Conti as president; Corsini and EAT representatives Guido Del Beccaro and Carlo Delcroix as vice-presidents; Garbasso as Government Commissioner; Gino Bargagli Petrucci, Carlo Martini Bernardi, Roberto Venturi Ginori, and Barfucci as councilors. The Executive Committee appointed several sub-commissions, each of them presided by experts in particular areas of science and members of Florence's scientific institutions.

5 On these topics, see Palla, *Firenze nel regime fascista*, 231–292.

6 Vincenzo Gioberti, *Del primato morale e civile degli italiani per Vincenzo Gioberti. Seconda edizione corretta e accresciuta dall'autore coll'aggiunta di una nuova avvertenza* (Brusselle: dalle Stampe di Meline, Cans e Compagnia, 1845), 501.

SUBCOMMISSIONS AND COMMISSIONERS

AGRICULTURE AND WATER MANAGEMENT

Giovanni Bellincioni (Counselor for Public Works of the City of Florence)
Giuliano Gondi (*Società Boracifera di Larderello*)
Arrigo Serpieri (president of the Georgofili Academy, Florence)

ANTHROPOLOGY AND ETHNOGRAPHY

Aldobrandino Mochi (director of the National Museum of Anthropology and Ethnography, Florence)

ASTRONOMY AND PHYSICS

Giorgio Abetti (director of the Arcetri Astronomical Observatory, Arcetri)
Giuseppe Boffito (Collegio alla Querce, Florence)

BOTANY

Giovanni Negri (director of the Botanical Garden of Florence)

CHEMISTRY

Luigi Rolla (University of Florence)

GEODESY AND GEOGRAPHY

Renato Biasutti (University of Florence)
Nicola Vacchelli (director of the Military Geographic Institute, Florence)

MEDICINE

Andrea Corsini (director of the Institute of History of Science, Florence)

SEISMOLOGY

Guido Alfani (director of the Ximenian Observatory, Florence)

TECHNOLOGY

Luigi Pasqualini (General Manager of the *Officine Galileo*)
Umberto Picchi (president of the National Association of Engineers, section of Florence)

ZOOLOGY

Vincenzo Baldasseroni (*La Specola* Museum, Florence)

MUSEUMS AND GALLERIES

Antonio Minto (University of Florence, Superintendent of Antiquities of Etruria)
Luigi Pernier (University of Florence)
Giovanni Poggi (director of the State-Owned Galleries of Florence)

LIBRARIES AND ARCHIVES

Angelo Bruschi (director of the National Library, Florence)
Umberto Dorini (*La Colombaria* Academy, Florence)

It is telling that all the people involved in the organization of the event were based in Florence. The idea of bringing in experts from outside the city was not even taken into consideration, as the main assumption of the whole enterprise was the identity between the Florentine/Galilean tradition and the Italian scientific tradition. The Executive Committee's composition also gives us the chance to see the patronage network that extended beyond the whole initiative. Vice-president Guido Del Beccaro was a lawyer and businessman who was involved in various capacities in fascist administration. Apart from being a member of the EAT board, he was president of *Autostrade Toscane* (Tuscan Highways Company), attorney for the City of Florence, and Ginori Conti's personal lawyer. He would be also involved in the creation of the *Maggio musicale fiorentino* and – together again with Ginori Conti and Garbasso – of the *Fiera internazionale dell'Artigianato* (International Handicraft Fair). The other vice-president, Delcroix, also a member of the EAT board, was a veteran of the Great War, a member of the Italian Parliament and the president of the hugely influential *Associazione Nazionale fra Mutilati e Invalidi di Guerra* (National Association of the Seriously Disabled or Mutilated at War – ANMIG). Roberto Venturi Ginori, finally, was a member of the narrow circle of Florentine social and economic elite which fascism had managed to take to its side by dispensing public offices and creating *ad hoc* business opportunities.

Securing the support of local fascist patronage network was only the first step towards the realization of a national event. The second and more important one was obtaining Mussolini's public and explicit approval. Knowing beforehand that they would get it only if they could demonstrate the propaganda value of the initiative, organizers offered Mussolini the Honorary Presidency of the Exhibition, and depicted the event as a spectacular affirmation of the Italian genius – a genius which fascism both guarded and fulfilled – and as an occasion to lay claims to Italian scientific priority. "This Exhibition of ours," the Committee wrote in the request sent to Mussolini at the beginning of May 1927, "is going to be a display of Italianity in the highest sense of the word."[7] According to the organizers, the Exhibition was to be held from March to June, 1928, and artifacts would have been put on display in disciplinary arrangement. "Within it," they pointed out, "technology will perform a lofty task, that is, revealing all the stages through which an instrument has gradually passed until it reached its latest refinement."[8]

They were hitting the nail on the head. In 1927, fascism was consolidating popular consensus by presenting itself as the restorator of the 'true' Italian spirit in opposition to the 'debauched and chatty Parliamentarism' of the past, and both history and science featured prominently in this program. Endorsing the idea of nurturing and popularizing "national science" already set forth by the Italian Society for the Advancement of Science (SIPS) in 1925,[9] the regime favoured and sustained celebratory initiatives such as those in honour of Stanislao Cannizzaro (Palermo, 1926) and Alessandro Volta (Como, 1927). The Exhibition of History of Science in Florence was not meant to celebrate an individual, but rather the Italian science as a whole, and to Mussolini this looked like a great opportunity to spread the image of fascism as the legitimate heir of the Italian scientific genius. So, on May 19, 1927, the *Duce* wired the following message to Garbasso:

> Honourable Senator and *Podestà* of Florence,
>
> 1° I fully approve of the "I° National Exhibition of History of Science" to be held in Florence from March to June, 1928.
>
> 2° I accept the honorary presidency.

---

7 "L'esposizione nazionale di storia della scienza," *Unità Cattolica*, August 6, 1927.

8 "L'esposizione nazionale," *Unità Cattolica*, cit.

9 See Monica Cioli, "Un congresso internazionale di fisica: l'organizzazione della scienza nella costruzione dello Stato fascista," in *Fascismo e scienza. Le celebrazioni voltiane e il Congresso internazionale dei Fisici del 1927*, ed. by. Aldo Gamba, Pierangelo Schiera (Bologna: il Mulino, 2005), 27–121: 89–91.

> 3° The Government will appropriately subsidize the enterprise, which is unprecedented, stupendous, and must turn out to be worthy of Florence, and of Fascist Italy.
>
> MUSSOLINI[10]

This brief message, having been greatly publicized by the Italian press, gave a terrific boost to the organizers' ambitions. Mussolini himself was acknowledging that their initiative was "unprecedented and stupendous," and so they could now turn to Universities, museums, and other institutions and urge them to contribute to the Exhibition. They even asked the National Institute of History of Science to officially recognize the initiative and get involved.[11] It is important to stress, at this point, that the regime's involvement did not obliterate the reductionist local/regional narrative embedded in the original plan. As we will see in the next chapter, fascism's approach to cultural enterprises allowed for the coexistence of local and national narratives without addressing the issue of their articulation, and so conflicting local traditions could be pushed even in supposedly 'national' cultural events. This turns out to be a factor of critical import not only in the designing of the Exhibition itself, but in the developing of a nation-building narrative within Italian museums in general.

Once organizers received Mussolini's approval, they called a meeting of the Executive Committee to be held on June 5, 1927, at the EAT headquarters. The Committee was to deal with issues related to the Exhibition funding, location, and organization, and hoped to quickly solve them. Things, however, would soon prove to be more complex than expected, and would force the Committee to postpone the Exhibition until 1929.

2 Funding and Financing

The very first problem to deal with was, quite obviously, that of funding. It was clear right from the start that the costs of the Exhibition could not be covered by the City of Florence and the EAT alone – not to mention the Institute of History of Science. According to the budget plan, the Exhibition would have

10 "Lettera di B. Mussolini al podestà di Firenze A. Garbasso: il capo del Governo approva l'iniziativa e accetta la presidenza onoraria dell'Esposizione, Roma 19 maggio 1927," ArchMG, *Esp29 – Organizzazione e allestimento, 1927–1929*, n. 3.

11 "Istituto Nazionale di Storia della Scienza. Comunicazioni ufficiali," *Archeion*, 8 (1927): 228.

costed no fewer than 500.000£, and thus a steady financial support from the Government and, possibly, from banks and industrial firms was needed.[12] Therefore, on June 1, 1927, Ginori Conti travelled to Rome to seek patronage from King Vittorio Emanuele III and speak to Mussolini in person about the "appropriate subsidy" the Committee had been promised.[13] He reported to the Executive Committee about this consultation during a meeting held on June 5. The *Duce*, he announced triumphantly, had decided to donate 300.000£ for the Exhibition and to allocate 400.000£ for the completion of the works on the ground floor of the National Library's new building in order to stage the Exhibition there. Moreover, during that same committee meeting Garbasso confirmed that the City of Florence was to give 100.000£, and the Executive Committee's secretary, Raffaello Bacci, informed that all of the paperwork required for obtaining official authorization for the Exhibition had been filled out and sent to the Ministry of Internal Affairs.[14] At the beginning of June, thus, things looked rather favorable.

This scenario dramatically darkened in August, when the Ministry of Finances informed the Committee that the promised 400.000£ could not be given at the moment. Since there were still no news about Mussolini's and Garbasso's announced contributions either,[15] the Exhibition's coffers were completely empty, out-of-pocket expenses being covered by Ginori Conti and the EAT. Meanwhile, the Committee had received the official authorization from the Ministry of Internal Affairs, and was waiting for the Ministry of Public Works' approbation of discounts on train fares for the Exhibition. Still, given the financial situation, the Committee could do very little beyond sending circular letters to Universities and Museums announcing the event. It was a troubling situation. As Corsini explained to Ginori Conti, the Exhibition had been advertised with great pomp in newspapers; it had been approved and was formally supported by Mussolini himself; boosted by Mussolini's blessing, several local committees involving politicians, scientists, historians, and men of learning had gathered together; the King had given his patronage, and the Committee had already made arrangements with contractors for

12 Raffaello Bacci to Piero Ginori Conti, May 29, 1927, ASF, PGC archive, s. 2, f. "Storia della Scienza (2)."

13 Piero Ginori Conti to Raffaello Bacci, May 31, 1927, ASF, PGC archive, s. 2, f. "Storia della Scienza (2)."

14 Raffaello Bacci to Piero Ginori Conti, July 13, 1927, ASF, PGC archive, s. 2, f. "Storia della Scienza (2)."

15 Andrea Corsini to Piero Ginori Conti, August 13, 1927, ASF, PGC archive, s. 2, f. "Storia della Scienza (2)."

construction works and furniture. In short, the Exhibition 'had' to be staged, no matter how, for failing would have been disastrous in terms of image and reputation. So, it was necessary, according to Corsini, to solicit the ministries of Finances and of Public Works about the money promised by Mussolini, and, in case, ask the *Duce* himself. Beyond that, it was also apparent that the Exhibition could not be staged in 1928. The financial situation was shaky, Corsini noted, and there were unresolved issues still waiting to be settled.[16] He straightforwardly explained all this to Ginori Conti on November 2, 1927.

> Money. We don't have a cent. Y[our] E[xcellence] knows very well that the 300.000£ promised by Mussolini became 80.000£ as Volpi [*Giuseppe Volpi, Minister of Finances*] took delivery of the allocation, and maybe they will become even fewer the day De Bellis [*Vitantonio De Bellis, State General Accountant*] records it. Thus, it is necessary to actively press the Government for the money, for it will surely give them, but not so soon. The City of Florence has promised us 100.000£; still, when we asked Senator Garbasso to give us at least a part of them within the present year, he told us that it will impossible for him to give any money until the City's budget for 1928 is approved.
>
> Subscriptions. Bacci, together with Del Beccaro, has been conducting a subscription campaign sending letters all over Italy for ten days now. Feedbacks are not pleasing so far, because the only institutions that answered are the Chamber of Commerce of Turin and that of Carrara, and they gave us 2000£ and 1000£, respectively, while the others told us that they will consider our requests in 1928, maybe. We raised no money at all from privates.[17]

Given such circumstances, Corsini suggested to postpone the Exhibition until 1929, so as to have more time to raise money, find a suitable location for the event, and thoroughly discuss the Exhibition arrangement. He reaffirmed this opinion in front of Ginori Conti, Garbasso, Del Beccaro, Barfucci, and Bacci in a meeting held on November 5 at Ginori Conti's house. Upon hearing Corsini's account of the situation, they unanimously agreed to postpone the Exhibition until 1929 and to stop advertising it in newspapers. It was also agreed that

---

16 Andrea Corsini to Piero Ginori Conti, August 13, 1927, ASF, PGC archive, s. 2, f. "Storia della Scienza (2)."

17 Andrea Corsini to Piero Ginori Conti, November 2, 1927, ASF, PGC archive, s. 2, f. "Storia della Scienza (2)."

Corsini and Bacci were to work on the arrangement, while Ginori Conti and Garbasso would continue to press on the Government for the money.[18]

So, in January, 1928, Ginori Conti travelled to Rome again to meet with Mussolini. This time, Garbasso and the Prefect of Florence, Giuseppe Regard, accompanied him. Mussolini received them kindly and had the Ministry of Finances allocate 380.000£ for the Exhibition on the spot.[19] The financial situation soon began to clear up, as the Chamber of Commerce of Florence had announced in December 1927 that it would contribute 20.000£, and similar announcements had been made by the chambers of Commerce of Arezzo (1000£) and Rome (5000£).[20] In May 1928, the City of Florence gave 100.000£, plus 50.000£ in December and 20.000£ in February 1929.[21] At that time, the total amount of money raised by the Executive Committee counted up to 617.716£: nearly 20.000£ higher than expected in the budget plan.[22]

CONTRIBUTIONS FOR THE EXHIBITION

FROM THE ITALIAN GOVERNMENT...380.000£

FROM THE CITY OF FLORENCE...170.000£

FROM THE CHAMBERS OF COMMERCE...29.000£

FROM PIERO GINORI CONTI...8000£

– **From subscriptions and donations...30.716£**

TOTAL: 617.716£

18 Raffaello Bacci to Piero Ginori Conti, December 1, 1927, ASF, PGC archive, s. 2, f. "Storia della Scienza (2)."

19 Piero Ginori Conti to Andrea Corsini, January 26, 1928, ASF, PGC archive, s. 2, f. "Storia della Scienza (2)."

20 Raffaello Bacci to Piero Ginori Conti, December 1, 1927, ASF, PGC archive, s. 2, f. "Storia della Scienza (2)."

21 ASCFi, CF 898, May 4, 1928; CF 903, December 4, 1928; CF 904, February 15, 1929.

22 "Financial statement. December 1930," ASCFi, CF 9260, s. 7.5, 371–425.

3 Location

Finding a proper location for the Exhibition was the second main issue that decisively contributed to the decision of postponing the event to 1929. As we have seen, Ginori Conti, during his first meeting with Mussolini, had received assurance that the Government would have taken charge of the completion of the ground floor of the National Library's new building in the Piazza dei Cavalleggeri allocating 400.000£ to this purpose. Organizers were confident that this would have worked out for the best, since the completion of the building was a work that the Government planned to do in any case. They were also confident that, after the Exhibition, they could set up the yearned-for National Museum of History of Science right there. So, at the beginning of August, 1927, they got in touch with an architect and charged him with the design of the exhibition space. According to the architect's sketch, the Exhibition would have consisted of five huge rooms, each devoted to a thematic area, that is: Natural Sciences; Medicine and Pharmacy; Mathematics, Physics, and Chemistry; Astronomy and Geography; Technology. This proposal met with the approval of the Executive Committee, but the bad news coming in from Rome around the middle of August impeded its realization.[23]

At this point, the Committee had two remaining options to choose from. The first one, favoured by Garbasso, was the Pitti Palace, for Garbasso himself had already made arrangements with Minister of Instruction, Pietro Fedele, to have some of its rooms at the Committee's disposal. The second one, favoured by Corsini, was the Palace of Expositions at the San Gallo *Parterre*, property of the City of Florence, even though it would have been available only in 1929. In his letter to Ginori Conti of November 2, Corsini so argued:

> Senator Garbasso, as I have written, says that Minister Fedele would grant us the Pitti Palace. Please allow me to analyze with you what kind of spaces we could have at our disposal in the Pitti Palace. On the ground floor there are no room for us, because the rooms are all already occupied by the *Museo degli Argenti* and certainly it would not be possible to move that museum elsewhere. [...] On the first floor, apart from the *Galleria Palatina*'s rooms, there are eighteen or maybe twenty rooms, but all of them, due to the lavishness and richness of their decorations, are

23 Andrea Corsini to Piero Ginori Conti, July 25, 1927; Andrea Corsini to Piero Ginori Conti, August 2, 1927; Andrea Corsini to Piero Ginori Conti, August 6, 1927, ASF, PGC archive, s. 2, f. "Storia della Scienza (2)."

> unfitting for an exhibition like our own, since we could affix nothing to the walls nor we could make use of showcases which, by covering frescoes and tapestries, would distort the whole ensemble of those beautiful rooms. On the second floor there are the Modern Gallery – which has been recently renovated at great expense – and the King's private rooms. In conclusion, I think that the Pitti Palace is unfitting and inadequate for our aims.[24]

With Ginori Conti's backing, Corsini managed to bring the Executive Committee to his side and the decision of holding the Exhibition in 1929 at the Palace of Expositions at the San Gallo *Parterre* was ratified during the meeting of November 5, 1927.

Originally, the San Gallo *Parterre* was a small garden set up just on the edge of the city center, on the area where the namesake Abbey, demolished in 1529, once stood.[25] It had been designed upon Grand Duke Peter Leopold's order by architect Anastasio Anastagi, and in the 18th century it was one of the Florentines' favourite places to stroll about or have picnics. Yet, by the early 19th century, it had begun to decay as the larger and more comfortable *Cascine* Park had opened to the public.[26] In 1865, Giuseppe Poggi's urban planning scheme integrated the *Parterre* in the city's urban fabric, raising the problem of its requalification.[27] Indeed, with the demolition of the Arnolfian walls, the transformation of the open space in front of Porta San Gallo into a monumental square centered on Jadot's triumphal arch, and the edification of a new medium-high end neighbourhood called *Le Cure*, the *Parterre* found itself within one of the city's prestigious new areas. So, in 1914, following a lively debate, it was decided to erect a Palace of Exposition "adequate for the city's needs, but by all means not sumptuous," at a maximum expense of 800.000£.[28]

The decision to provide the city with a venue for exhibiting modern and contemporary art fitted in with the already mentioned institutional activity of promoting cultural tourism. It was the *Associazione per il movimento dei forestieri* of Florence (Association for the Movement of Foreigneirs), a national

---

24 Andrea Corsini to Piero Ginori Conti, November 2, 1927, ASF, PGC archive, s. 2, f. "Storia della Scienza (2)."

25 See Alfred von Reumont, *Tavole cronologiche e sincrone della storia fiorentina compilate da Alfredo Reumont* (Firenze: G.P. Viesseux, 1841), v, 28.

26 On the *Parterre*, see Mario Bencivenni, Massimo De Vico Fallani, *Giardini pubblici a Firenze dall'Ottocento a oggi* (Firenze: Edifir, 1998), 115–124.

27 See Carlo Cresti, *Firenze, capitale mancata: architettura e città dal piano Poggi a oggi* (Milano: Electa, 1995), 9–49.

28 Quoted in Bencivenni, De Vico Fallani, *Giardini pubblici a Firenze*, cit., 123 note 2.

society intent upon promoting tourism founded between 1898 and 1899, which raised the money for the construction of the Palace and appointed architect Enrico Dante Fantappiè and engineer Vittorio Tognetti to design it. The two of them created a huge eclectically neoclassical building made out of field-stones, covered with an Eternit roof held up by reinforced concrete columns and sustained on iron trusses.[29] The entire structure was composed of a main body of pentagonal layout built on three levels (two above ground plus a basement), and of a secondary, single-storey body of rectangular layout. The main entrance consisted of a three-arches portico, on whose architrave stood out the inscription "For our life the Art." The ground floor of the main building's interiors featured two large central Salons encircled by smaller rooms, a layout characterizing also the basement and the secondary building.

The Palace of Expositions was inaugurated on April 8, 1922, on the occasion of the first edition of the Painting and Sculpture Exhibition *La Fiorentina Primaverile* (Florentine Spring Exhibition of Painting and Sculpture).[30] Over the years, it housed important cultural events such as the Florence International Book Fair (prompted by Barfucci), and the Annual International Handicraft Fair.[31] The Palace lasted until 1937, as in that year the City of Florence charged architects Sirio Pastorini and Mario Pellegrini with the task of renovating the *Parterre*. The Palace of Expositions was demolished and the outdoor area completely reshaped. At the end of this renovation, the *Parterre* changed its name into *Giardino dei Littoriali* (the *Littoriali*'s Garden – *Littoriali* was the Fascist name for University Sports, Arts, Culture, and Working Contests).[32]

On the occasion of the Exhibition of History of Science, three small pavilions were erected along the path leading to the Palace. They were built to house the Ministerial exhibitions,[33] and their construction was explicitly required by the Ministries themselves. When the matter came up before the Executive Committee, Del Beccaro had opposed the Ministries' request arguing that, had the Committee agreed to it, expenses would have been far higher than

29 Emanuela Piccini, "Opere e progetti di Enrico Dante Fantappiè (1869–1951). Indagini e osservazioni per la conoscenza e il restauro," *bollettino ingegneri*, 7 (2007), 18–26: 20–21.

30 See *La Fiorentina Primaverile: prima esposizione dell'opera e del lavoro d'arte nel palazzo del Parco di San Gallo a Firenze, 1922. Catalogo delle opere esposte, con cenni biografici e critici* (Roma: Casa d'arte Valori Plastici, 1922).

31 On these initiatives, see Carla Campisano, "La Fiera Internazionale del Libro di Firenze, 1922–1932," *Bollettino di storia dell'editoria in Italia*, XIII/1 (2007), 19–22; Tina De Rocchi Storai, "La Mostra Internazionale dell'Artigianato di Firenze," *Arti e Mercature*, IX, 9 (1972).

32 See Gabriele Corsani, "Tre Piazze dell'Ottocento a Firenze," in *Storia della città. Le Piazze. Lo spazio pubblico dal medioevo all'età contemporanea*, 54–55–56 (1990), 165–174: 171.

33 On Chiaramonti's buildings, see Carlo Cresti, *Architettura e fascismo* (Firenze: Vallecchi, 1986), 239–240.

expected in the budget plan. To this, Ginori Conti answered that they "cannot do otherwise," and that he himself "not even had the time to call a meeting to discuss the matter."[34] The three pavilions joined the two already built on the occasion of the International Book Fair of 1927; so, in total, the Committee had at its disposal an exhibition area consisting of 46 rooms distributed among 5 pavilions, with a total surface area (including outdoor spaces) of about 2500 square meters.

4 Exhibition Criteria and Local Committees

As stated in the general plan of May 1927, the Exhibition, in order to best represent the Italian contribution to science, needed the support and cooperation of the whole country. Therefore, the Executive Committee, wishing for the establishment of local committees, had already begun sending circular letters announcing the event to Universities, Museums, Libraries, High Schools, and Scientific Institutions. At first, reactions to the news of an exhibition of history of science in Florence were rather cold, but after the publication of Mussolini's letter of approval local committees began to be established all over Italy. At the end of May, local committees existed in Rome, Bologna, Brescia, Milan, Livorno, Lucca, Bergamo, and Naples, while others were being established elsewhere.[35] At the beginning of June, the Executive Committee issued a circular letter outlining their role. The local committees were supposed to report about anything relating to history of science – that is, "*portraits, autographs, relics* of Italian scientists and travelers; *scientific artifacts, books, treatises, plates etc.* of historical relevance or somehow related to the history of science" – they happened to have news of.[36] Then, upon receiving each local committee's list of items, the Florentine commissioners would have sorted them out according to the Exhibition criteria.

According to the original plan, the Exhibition would have been structured into five sections corresponding to five areas of science:

I. Natural Sciences;
II. Medicine and Pharmacy;
III. Mathematics, Physics, and Chemistry;

---

34 Guido Del Beccaro to Piero Ginori Conti, March 23, 1929; Piero Ginori Conti to Guido Del Beccaro, March 23, 1929, ASF, PGC archive, s. 2, f. "Storia della Scienza (1)."

35 For detailed informations about the local committees, see below, part II.

36 "Circolare-programma," in *Guida della Prima Esposizione Nazionale di Storia della Scienza* (Firenze: Ente per le Attività Toscane, 1929), 20. Italic in the original.

IV. Astronomy and Geography;
V. Technology.[37]

This arrangement, the organizers thought, allowed them to resoundingly demonstrate how Italy had been "the world's teacher in every field of science," and, through the exhibition of the progressive refinement of scientific instruments, hint at the idea of the "living museum" which fascinated Mussolini so much.[38] So, they specified in the circular letter sent to the local committees, the Executive Committee would take care of the set-up of each room and, thanks to the local committees' work, would compile a general catalogue of the Italian scientific heritage as an "unassailable proof" of the Italian scientific primacy.[39] What the Executive Committee asked them to do, then, was 'just' to locate, collect, classify, and describe – according to a set of rules provided by the Executive Committee – the historical scientific collections held in their cities, and, moreover, to provide bio-bibliographical information about the scientists evoked in their reports. Considering that a large part of the Italian scientific collections were still waiting to be reviewed and inventoried in the first place, these were exceedingly arduous tasks.

In November, just a few of the contacted committees had replied – in fact, some of them had got in touch with the organizers just to let them know that, if the Exhibition was to be held in 1928, they could not participate. In September 1928 the meeting of the Italian Mathematicians would indeed be held in Bologna and an historical Exhibition celebrating the 400th anniversary of the birth of King Emanuele Filiberto di Savoia and the 10th of the victory in the Great War was scheduled to be set up in Turin around the same time. Thus, representatives from those two cities communicated that they could not undertake any action for the Florentine exhibition. Solicitations sent to other local committees mostly fell on deaf ears.[40]

Apart from the huge amount of work that the Executive Committee demanded from the local committees – very likely a work to be carried out in precarious conditions, without any economic support, and which would not bring immediate benefits, except maybe the organizers' gratitude – there were another reason that warned everyone to be cautious about being involved in the Exhibition. From a confidential letter sent by Ginori Conti to the presidents of

37 "Circolare-programma," in *Guida della Prima Esposizione Nazionale di Storia della Scienza*, cit., 18.

38 See Marco Beretta, "Oggetti, spazi e idee: la storia materiale della scienza," in *Scienze e storia nell'Italia del Novecento*, ed. by Claudio Pogliano (Pisa: Plus, 2007), 157–173: 165–167.

39 On this topic see below, chapter 5.

40 Raffaello Bacci to Piero Ginori Conti, November 29, 1927, ASF, PGC archive, s. 2, f. "Storia della Scienza (2)."

the local committees in December 1927, it appears that rumours were spreading that the actual goal of the Exhibition was to give life to a National Museum of History of Science and thus that the objects sent to Florence would not be returned. Understandably enough, this had caused worries and concerns on the local committees' part, and it seems that some of them had decided to withdraw their participation unless they were provided with appropriate guarantees. First of all, they wanted the Executive Committee to make a formal commitment to give back all the borrowed objects; secondly, they asked that tags noticeably indicating the artifact's place of origin and the name of its owner be put on each object; thirdly, they proposed that the Exhibition be organized geographically, not thematically, in order to avoid scattering collections across several rooms; finally, they claimed the right to select what to put on display and to curate their own exhibits.[41]

It is impossible to positively assert which one of the committees raised the issue because the rumours were mostly spread by word-of-mouth and no direct documentation of the committees' requests is currently available. Yet there were committees that had reasons to be wary, and one of them surely was the Roman committee, which featured representatives of the Royal Observatory of the Campidoglio who at the time were dealing with the open question of the Copernican Museum, whose collections, after the suppression of the Observatory of the Roman College, were temporarily entrusted to the Lyncean Academy. In 1923, Guido Horn d'Arturo had already tried to have the Museum moved to Bologna;[42] a year later, as seen above, Mieli had advanced the idea of assigning it to the Federation; and at the turn of the decade Corsini and Ginori Conti would make similar attempts relying, as we shall see, on the claim that, "while the Museum holds almost nothing about Copernicus himself, there are many artifacts and objects directly linked, because of their authors and their history, to the Galilean relics".[43] Perfectly aware of the situation, on April 20, 1929, Giuseppe Armellini, director of the Observatory of the Campidoglio and member of the Roman committee, published an article titled *Roma, Firenze e l'astronomia siderale* ("Rome, Florence, and Sidereal Astronomy") on "il Giornale d'Italia." In it, Armellini argued against the

41 Circular letter, November 20, 1927; Circular letter [s.d.], December 1927, ASF, PGC archive, s. 2, f. "Storia della Scienza (2)."

42 See Fabrizio Bònoli, "Il Museo della Specola del Dipartimento di Astronomia e del Sistema Museale d'Ateneo dell'*Alma Mater Studiorum*," *Annali di Storia delle Università italiane*, 17 (2007), 489–497.

43 See below, chapter 6.

identification of the Italian scientific tradition with the Galilean/Florentine one, stressing on the existence of a genuine "Roman" tradition.

This controversy is important not only because it threatened to thwart the entire Exhibition, but mainly because it shows how difficult it was to negotiate between local and national narratives, and how instruments and collections were still understood primarily as constituents of local identities. Though acknowledging that the groundwork for the birth of modern astrophysics had been laid in Florence, Armellini maintained that it was in Rome that the discipline fully developed. In 1869, the Observatory of the Campidoglio, directed by Lorenzo Respighi, and that of the Roman College, directed by Angelo Secchi, were two of the world's leading institutions in the spectroscopy research field. "At the beginning of 1869," Armellini pointed out, referring to Lorenzo Respighi's objective prism, "only the observatories of the two Roman universities owned this new and great instrument of inquiry, and so they forerun all the others in the astrophysics field."[44] Though delivered in a conciliatory tone, the message was clear: Secchi's, Respighi's, and others' relics, which constituted part of the Copernican Museum's collections, vouched for the existence of an "essentially Roman" tradition, and so they had to stay in Rome.

As it would turn out, rumours were not far from being true, for Ginori Conti, Corsini, and Garbasso actually hoped to convert the Exhibition into a permanent Museum. Just before the closing of the event, Ginori Conti would send thank-you letters to the local committees, tentatively asking them to lend their collections to the new history of science museum that the Executive Committee was trying to set up (needless to say, no one would acquiesce to the request).

> Since amongst the materials sent to the exposition there are some that, as they are kept isolated, do not constitute 'documentation', and since the same would acquire fundamental importance if held in a proper institute, we are now asking the said collections, instead of being sent back to their owners, be kept in Florence as part of the new Museum of history of science.[45]

In December 1927, however, the organizers accepted the conditions set forth by the local committees: they committed themselves to return all the borrowed

---

44 Giuseppe Armellini, "Roma, Firenze e l'astronomia siderale," *il Giornale d'Italia*, April 20, 1929.

45 Circular letter [s.d] October 1929, ASF, PGC archive, s. 2, f. "Storia della Scienza (2)."

artifacts to their owners, to give the local committees the opportunity to select what to put on display, and, above all, to reshape the Exhibition as a "collection of regional expositions."[46]

The latter decision, in particular, carried important consequences. First of all, it was impossible to exhibit "the progressive refinement" of scientific instruments and apparatuses, which should have been the focal point of the Technology section – the five sections themselves making no sense anymore, now that the local committees had been given the opportunity to manage their own exhibits. The geographical order, moreover, did not allow to show and celebrate the Italian contribution to science as a whole: more than a celebration of Italian science, the Exhibition would eventually turn out to be a celebration of local traditions. Finally, lacking an overall thematic ordering, the Exhibition could not pursue serious didactic ambitions. By adopting a geographical ordering and letting each local committee manage its own show, organizers turned a thematic Exhibition into something more similar to a Universal Exposition – an eclectic, unregulated "jumble of foreignness."[47] It is also interesting to note how local Italian traditions acted as national traditions, creating the same tension between thematic and geographical ordering that characterized universal expositions since their earliest editions.[48] On the bright side, staging a "collection of regional expositions" required less time and money while showing more clearly Florence's supposed primacy. The Exhibition would indeed draw a 'topography' of Italian science, exhorting visitors to make a comparison between the featured cities. Recreating the typical competitive mood of Universal Expositions, it would build the ideal environment for nurturing and promoting the organizers' reductionist view, substantiating the claim that the local Florentine scientific tradition was essentially identical with the Italian one.

46 "L'Esposizione di Storia della Scienza." Interview with Andrea Corsini, *il Giornale d'Italia*, July 28, 1928.

47 See Meg Anderson, "'A Jumble of Foreigness': the Sublime Musayums of Nineteenth-Century Fairs and Exhibitions," *Cultural Critique* 23, 1992, 199–250.

48 Bruno Giberti, *Designing the Centennial: a History of the 1876 International Exhibition in Philadelphia* (Lexington: the University of Kentucky Press, 2002).

CHAPTER 4

The Exhibition in Context

1 History of Science and Propaganda: The Italian Genius and the National Scientific Primacy

The First National Exhibition of History of Science officially opened on May 8, 1929, at about 10.00 AM. The ceremony was given widespread coverage in the media. King Vittorio Emanuele III, patron of the event, arrived in Florence by train early in the morning. Upon his arrival at the *Maria Antonia* train station (now *Leopolda*), he was greeted by the *podestà* of Florence, Giuseppe Della Gherardesca, the undersecretary to the president of Council, Francesco Giunta, the Cardinal Alfonso Maria Mistrangelo, and the Minister of National Economy, Alessandro Martelli. After a brief stop at Palazzo della Signoria, the King was driven to the *Parterre*. A short *Istituto LUCE* newsreel shows the diminutive Vittorio Emanuele III in military outfit and accompanied by a bevy of dignitaries and military officers as he passes through the garden and shakes hands with Ginori Conti, Corsini, and Garbasso at the main entrance of the Palace of Expositions in front of a large exultant crowd. *Carabinieri* in dress uniform and trumpeters in faux-Renaissance costume welcome the sovereign as fascist salutes are munificently exchanged.[1]

If the Exhibition succeeded in pointing out the importance of historical scientific collections as 'cultural objects', it was mainly because its organizers managed to make its aims consistent, at least *prima facie*, with those of the fascist regime.[2] The actual existence of a 'fascist ideology' has long been a controversial issue among historians. According to many, if one were to define 'ideology' as a coherent set of ideas and theories that serve as theoretical ground for action and mutual acknowledgement, then there is no such thing as a 'fascist ideology'. Fascism, according to this view, endorsed different ideologies on a case-by-case basis, mostly reusing pre-existing ones as a mere justification for its daily, everchanging political practice. Fascism, has also been said, was a political movement with little or no culture at all. This belief is in

1 The newsreel is available at the website http://www.archivioluce.com/archivio/jsp/schede/videoPlayer.jsp?tipologia=&id=&physDoc=1274&db=cinematograficoCINEGIORNALI&findIt=false&section=/.

2 About this topic, see, among others, Carlo Bordoni, *Cultura e propaganda nell'Italia fascista* (Messina-Firenze: D'anna, 1974); Philip V. Cannistraro, *La fabbrica del consenso: fascismo e mass media* (Roma-Bari: Laterza, 1975); Emilio Raffaele Papa, *Fascismo e cultura* (Venezia: Marsilio, 1975); Alessandra Tarquini, *Storia della cultura fascista* (Bologna: il Mulino, 2011).

 | DOI:10.1163/9789004297418_005

part inspired by fascism itself, since fascists took pain to present fascism as a movement committed to action, not theory, and resting upon facts, not ideas. As Mussolini put it, "Fascism makes history, does not write it."[3]

While fascism's strong commitment to action is undeniable, also undeniable is the fact that fascism did indeed have an ideology and was to some extent committed to theory and ideas. Even though its ideology was not as original, coherent, and logically sound as one might expect, it featured nevertheless a set of powerful ideas, reasonings, myths, and methods that allowed its supporters, at the very least, to recognize each other as 'fascists' and act accordingly. Fascist ideology mostly emerges in political practice and this is especially true in the case of events like the First National Exhibition of History of Science. It would be difficult to account for the interchange between cultural and political factors at play in such enterprises without referring them to some kind of fascist ideology. Events like the Exhibition are neither empty tools of propaganda nor unbiased cultural outputs, but rather complex political and cultural products enclosing a number of diverse ideas, practices, aims and premises.

During his early years as socialist militant, Mussolini developed views that stayed unchanged in the vortex of fascism's ever-changing policies. Loathing for democracy, aversion to masses, history as a place of struggle between elites were but a few of the ideas that helped fascism bring about its most fundamental aim: creating a new man, a new state, a new Italy. At first, during the early twenties, fascism's declared goal was to make people adhere to State. Socialism and socialists were seen and presented as fundamentally anti-Italian, so the first actions undertaken by the *squadristi* were directed against socialist cultural circles (*case del popolo*, *dopolavoristi* clubs, and the like). These actions were presented as 'Italian' in spirit, rather than 'fascist', and were not explicitly directed at creating a fascist regime. Early fascists would argue that what they wanted was just to reinforce a State on the verge of being disrupted by socialists and anarchists. This way, they gained wide support from conservatives. At the same time, the squadrists' actions were instrumental in bringing about the conviction that being fascist essentially coincided with being Italian. As fascism became popular and powerful, it became increasingly clear that "Italian" was to be considered a synonym for "fascist," and therefore that the fascist motto "new man, new State, new Italy" should be understood as "fascist man, fascist State, fascist Italy."[4] This was by all means a cultural operation, and so fascism put a great deal of effort into propaganda, especially in the aftermath

3 See Emilio Gentile, *Le origini dell'ideologia fascista: 1918–1925* (Bologna: il Mulino, 2011).

4 See Victoria de Grazia, *The Culture of Consent: Mass Organization of Leisure in Fascist Italy* (New York: Cambridge University Press, 2010).

of Matteotti's murder, in order to present itself in continuity with, rather than in opposition to, the Italian glorious past in general and the *Risorgimento* in particular. Even though fascism never set out a strict set of rules to define what 'fascist art', 'fascist architecture', and 'fascist culture' must be, it nonetheless prompted intellectuals to discuss such themes and equipped them with myths, ideas and imagery that were presented as inherently 'Italian' (that is, fascist). One of the most important among them was the 'Italian genius'.

The theme itself was hardly new. During the *Risorgimento*, the myth of the Italian Genius enjoyed widespread popularity and played a central role in the discussions about the best way to attain political unification. It was a shared opinion that the 'Genius' was a nation-specific, environment-dependent feature that determines people's general attitude towards reality. For champion of neoguelphism Vincenzo Gioberti, it was a "disposition of the Spirit;"[5] Giuseppe Ferrari, a philosopher and most prominent advocate of federalism, called it a "undefined inspiration."[6] "Italy is a religion," Giuseppe Mazzini wrote in 1866, five years after the unification. "And the fundamental character of the Italian Genius is religious."[7] From this common, religious-like understanding of the genius descended the most diverse political interpretations, according to how differently each author presented what the Italian Genius was like. If Gioberti considered it inseparable from Catholicism, Alfredo Oriani, a writer and historian held in high esteem by Mussolini (who in 1931 would curate the publication of Oriani's collected works), characterized it as "sceptic in religion, rationalist in philosophy, juridical in reforms, and universal in its aspirations."[8] The myth of the Italian Genius enjoyed great popularity amongst Italian scientists of the XIXth century who, well before Italy's political unification, looked at themselves as a national community and would routinely meet at the *Riunioni degli scienziati italiani*.[9] On a historiographical level, the myth functioned as a conceptual fixed point around which the very idea of a national scientific tradition was construed and promoted. The history of Italian science was presented as a succession of solitary individuals who, thanks to their inherent genius,

5 Vincenzo Gioberti, *Del primato morale e civile degli Italiani* (Bruxelles: Dalle stampe di Melina, Cans e Compagnia, 1844, terza edizione), vol. II, p. 272.

6 Giuseppe Ferrari, *Frammento sui neo-guelfi*, in *Opuscoli Politici e Letterari* (Capolago: Tipografia Elvetica, 1852), p. 223.

7 Giuseppe Mazzini, "Questione Morale," *il Dovere. Giornale politico settimanale per la democrazia* 4, issue 9, March 3, 1866.

8 Alfredo Oriani, *La lotta politica in Italia. Origini della lotta attuale (476–1887)* (Torino-Roma: L. Roux e C., 1892), p. 146.

9 See Marco Ciardi, "Scienza e Risorgimento Nazionale," in *Il contributo Italiano alla storia del pensiero – scienze* (Roma: Istituto della Enciclopedia Italiana, 2013), edited by Antonio Clericuzio and Saverio Ricci, pp. 351–357.

achieved great results in spite of political fragmentation and religious censorship. The *Tribuna di Galileo*, a scientific sanctuary inaugurated on the occasion of the third meeting of Italian scientists and celebrating Galileo almost like a Christian saint, is the most eloquent example of this approach.

Mussolini had been fascinated with the Italian genius since early in his career and elaborated on the theme throughout his rise to the top: writing in 1909, the 26-year-old journalist and political agitator stated that the Italian Genius was the most accomplished variety of genius and that Italy was the proper place for every genius to unfold itself.[10] Some twenty years later, as *Duce* of fascism and sole head of the Italian State, he urged historians to "trace back the footsteps of the Italian Genius" and made it the very theme of Italy's participation in the Universal exhibition of Chicago.[11] During the fascist era, the Italian genius was seen as the main factor behind Imperial Rome's greatness, Renaissance, and the rise of experimental science. In this ideological panorama saturated with ultranationalistic rhetoric, history of science was held up as the instrument for uncloaking the 'looters' who had 'seized' the Italian scientific glories throughout the centuries. Precisely in order to gain momentum from fascist rhetoric, the Florentine exhibition was alternatively presented by the Executive Committee as a "display of Italianity," a huge collection of "national priority claims," and a demonstration of how Italy had been "the world's science teacher." Even though Corsini would eventually be able to drive forward his genuine ideas about historical scientific collections, thus steering around from propagandistic suggestions, the Exhibition itself was nonetheless permeated with propaganda.

The Italian scientific primacy was the main theme of Minister Alessandro Martelli's inaugural speech. A stout man in his fifties, Martelli had been teaching Mineralogy and Geology at the University of Florence before entering the fascist party in the early 1920s. In 1926, he had been nominated as Undersecretary of Transports, and later, in 1928, as Minister of National Economy. Martelli was a strong supporter of the EAT – in fact, at that very moment, he was its president – and an impassioned advocate of Florentine cultural primacy.[12] In 1919, he had been amongst the promoters of the celebrations in honour of Leonardo da Vinci – considered as the prototype of the Tuscan genius – and had been working since then to create in Vinci a Museum and a Library devoted

10 Benito Mussolini, "Un grande amico dell'Italia: Augusto von Platen," in *Opera Omnia*, II, p. 171.

11 On the Italian participation at the Chicago World Fair see Giovanni Paoloni, Roberto Reali, Laura Ronzon, *I "primati" della scienza. Documentare ed esporre scienza e tecnica tra fascismo e dopoguerra* (Torino: Heopli, 2018.)

12 For a biographical account, see the entry "Martelli, Alessandro" in DBI, edited by Alessandra Tarquini.

to Leonardo. The Exhibition, Martelli explained in his speech, was a collection of "unassailable proofs" of the Italian scientific primacy. Being an "Italian" event "in the totalitarian sense of the word," it was heartily supported by the fascist regime, for "no form of thinking can be declared to be more Italian, than the scientific thinking." Italy, according to Martelli, had taught to the world the experimental method and invented applied science, and so "every foreign visitor of the Exhibition must experience a stronger-than-ever feeling of respect and admiration for the Italian people, which is valiant on battlefield, and brilliant and prolific in art, science, and labour." The same arguments were reiterated by Ginori Conti in his own speech. The Exhibition, Ginori Conti pointed out, had ignited historical researches which were to provide evidence of how "several inventions and discoveries hitherto attributed to foreigners are actually the product of the sole Italian genius."[13]

It was newspapers, however, that pushed these very arguments to their full extent. *Il Giornale d'Italia* presented the Exhibition as a "powerful affirmation of the Italian Genius;"[14] *il Popolo di Roma* praised the fascist regime for having restored national unity, thus paving the way for the advancement of science;[15] and *il Tevere* peremptorily stated that the Exhibition was a mighty demonstration that fascism had not "turned its back on science."[16] A massive press campaign was orchestrated by the Exhibition's press office manager, Dino Brogi. Born in Empoli in 1887, Brogi was a skilled propagandist, having already been in charge of the Florence Autonomous Agency for Tourism's press office.[17] He had also contributed to *Quaderni Fascisti* ("Fascist Notebooks"), a book series designed and edited by Piero Domenichelli, published by Enrico Bemporad's publishing company. *Quaderni Fascisti* had been devised in 1926 as "a collection of national propaganda addressed to the youth and the common people," with the explicit aim of backing up school teaching in the construction of political consensus.[18] Drawing inspiration from *Quaderni Fascisti*, Brogi conjured up a series of short stories that expanded on the myth of the Italian genius.

---

13 "Il Re ha inaugurato a Firenze la Mostra nazionale di storia della scienza," *il Giornale d'Italia*, May 9, 1929.

14 "Il Re ha inaugurato a Firenze la Mostra nazionale di storia della scienza," *il Giornale d'Italia*, May 9, 1929.

15 "Il Re inaugura solennemente a Firenze la mostra storica delle scienze e il congresso delle tradizioni popolari," *il Popolo di Roma*, May 9, 1929.

16 "Scienza e poesia," *il Tevere*, May 9, 1929.

17 On Brogi, see *Annuario della Stampa Italiana*, vol. 15–16 (Firenze: Casa Editrice del Libro Italiano, 1937), 229.

18 See Ugo Piscopio, *La scuola del regime: i libri di testo nelle scuole secondarie sotto il fascismo* (Napoli: Alfredo Guida Editore, 2006), 154–165; Carla Ida Salviati, "'Sor Enrico'. Ritratto di un grande editore," in *Paggi e Bemporad editori per la scuola. Libri per leggere, scrivere e far di conto*, edited by C.I. Salviati (Firenze: Giunti, 2007), 47–50.

The series was informally called "The primacy of Italian Science" and consisted of 16 stories whose mood and content is made explicit in their very titles: "The Italian Genius;" "Antonio Pacinotti's Glory;" "The Incandescent Lamp: an Italian Invention;" "How Nobel Took Advantage of an Italian Invention;" "An Heroic Scientist from the *Risorgimento*;" "The First Inventor of the Telephone;" "Harvey, Usurper;" "One truth amongst the impostures of Bell, the Bastard;" and so on.[19] In these stories, Brogi took to an extreme the idea of history of science as an instrument for laying claims to scientific priorities, polarizing the opposition between Italian scientists – depicted as brilliant inventors and discoverers or, at worst, ill-fated pioneers – and foreign "usurpers," in a rhetoric merging scientific chauvinism, nationalistic pride, and political propaganda.

> The field of history of science, untamed and abandoned for a long time, is now feeling the farmer's foot and Fascism's healthy plow. To tell the truth, even before this glorious and fruitful era some voices had been raised, hailing as a civil sacrilege every time a scientific glory of ours was presented in foreign make-up; but they were isolated, veiled voices, and anyway *clamantes in deserto* [...]. Not that it could have been otherwise. The old Italy of academic and foolish parliamentarism, the Italy that had witnessed with idiotic passiveness to the clamorous trial against Graham Bell filed in the U.S., [...] the chatty and blabbermouthed Italy that, failing to recognize the genius of Agostino Bassi, has granted Pasteur an undeserved glory; [...] that weak and jovial Italy, I say, [...] didn't have either time or mind to devote to such a beneficial matter as the reputation of our scientific glory. [...] Science stood at our Homeland's doorway, just like an unknown and vagabond bum begging for shelter.[20]

Stories like this began being published in January 1929 in both national and local newspapers. In order to move the public's collective imagination, Brogi emphasized the practical aspect of science and the social role of Italian scientists. They had never dodged their civic responsibilities by devoting themselves entirely to 'abstract' research, Brogi claimed. On the contrary, they had always worked bearing in mind the practical purpose of science. Precisely in this, according to Brogi, had consisted "the Italian genius": in the ability to

19 The other stories were "How the typewriter was born," "Sevres Porcelain and the Italian Genius," "At the Threshold of Mars," "Vannoccio Biringucci and Alchemists," "Italy: First in the World in Fighting Tuberculosis," "Our Share of Pasteur's Glory," "In the Field of Scientific Priority Claims," "Giovanni Cavalli, the Creator of Artillery."

20 "Nel campo delle rivendicazioni scientifiche," *Terra di Lavoro*, January 5, 1929.

direct abstract scientific research toward practical purposes for the benefit of the entire society. And it was this kind of Italian genius that fascism was now about to restore.

> Let the pure heirs of the scientific method's absolutism resign: science for itself – yes – but only at a first stage. Afterward, and immediately, science in its applications. [...] After all, Italy, following once again the Duce's command, is getting back on the track of its past glories, those same glories which Fascism wants to hold up as models of our breed. [...] We are now following the trail of the purest Italian genius.[21]

Of course, these were the very themes characterizing, in general, the regime's whole approach to science and history of science. Brogi, stressing on the idea of the "Italian tradition" and on the myth of the Italian genius as a mash-up of experimentalism and abstract research, was merely reformulating them in the aggressive and pompous language of propaganda.

2 History of Science and Ideology: The Italian Scientific Contributions throughout History

Besides being spread by newspapers, the idea of the "Italian scientific primacy" and the myth of the "Italian genius" associated with the Exhibition were given scholarly treatment as well. Two rounds of conferences, held between June and October, 1929, at the Palace of Expositions, were aimed at discussing "the everlasting contribution provided by so many Italians to the advancement of science."[22] They were organize by Garbasso, who invited several historians to present short *lectiones magistrales* illustrating the role of Italian scientists in the history of science.

The first session was opened on May 23 by Silvestro Baglioni, who presented a paper on "The historical development of Medical Biology in Italy," and went on with speeches by Guido Alfani ("Seismology in Italy"), Ettore Bortolotti ("Algebra in Italy from Leonardo Pisano to Paolo Ruffini"), Mariano Borgatti ("A History of Military Science"), Gaetano Scorza ("The Italian School of Geometry"), Giorgio Abetti ("The Italian Contribute to Astronomy and Astrophysics"), Antonio Garbasso ("The Italians' Contribute to Physics"), Davide Giordano ("A Look at the Italian Surgery from the Rome of the Lictorials

21 "Il genio italiano," *la Voce del Salento*, May 19, 1929.

22 "Un ciclo di conferenze alla Mostra di storia della scienza," *il Corriere d'Italia*, May 22, 1929.

to the Rome of the Lictorial"), and Agostino Gemelli ("The Relationships between Science and Philosophy in the History of Italian Thought").

The second session, held from September through October, should have featured Giacomo Acerbo, Giuseppe Belluzzo, Renato Biasuzzi, Filippo Bottazzi, Raffaello Nasini, and Guglielmo Marconi as speakers, but since many of them withdrew, it eventually featured Oreste Mattiolo ("A Look at the History of Botany in Italy"), Raffaele Giacomelli ("Flight in Nature and in Technology"), Eugenio De Vito ("At the Beginnings of Marine Engineering"), and Roberto Marcolongo ("Leonardo da Vinci's inventions"). Raffaello Nasini, though unable to attend the conference, promised to deliver a paper on "The Italian Chemistry" to be published in the proceedings. Still, he died in 1931 before accomplishing his work, and so Garbasso contacted Giulio Provenzal, who commemorated Nasini in a paper titled "Pages of History of Chemistry."

The proceedings of the conferences were published in 1932, and on that occasion the editor Gino Bargagli Petrucci recalled the event and its aims.

> In the huge Conference Room of the Exhibition, eminent Orators, invited by the Executive Committee, charged themselves with the task of recollecting many of those Names in the Papers now reunited in the present volume, so that not only Italians, but Foreigners too, could know and remember them; and remember above all that in many and many unexplored fields the first track has been Italian; a modest, silent, unselfish track, too often erased by the noisy shuffling of those who arrived later just to exult and, not infrequently, to … collect money….[23]

However, apart from the vindication of the Italians' scientific merits, the conferences aimed to demonstrate how the aptitude to scientific research was a distinctive feature of the Italian character and how fascism supported and promoted it. This was eloquently pointed out by Garbasso himself in his *lectio* on the history of Physics, a discipline that, being intent on discovering "the objective laws of reality" or "God's thought," was, according to him, "the archetype of science itself." Physics, for Garbasso, was a "characteristically Italian science," because "a profound sense of reality" is inherent in the Italian character.

> Let us approach with reverence the multitude of our physicists, then. Not only they have taught civilized people how to formulate and solve the problems of nature; not only they have created mechanics or electrology

23 Gino Bargagli Petrucci, "Introduzione," in *L'Italia e la scienza*, edited by G. Bargagli Petrucci (Firenze: Le Monnier, 1932), VI.

> and have designed some of the most beautiful machines; they had also been amongst the purest exponents of the Italian intelligence in general.[24]

From these premises, Garbasso deduced strong ethical and political judgements. For him, aversion to science and political opposition to fascism were but two sides of the same anti-national sentiment which animated people like Benedetto Croce, whom Garbasso accused of being the main responsible for the spreading of idealism and, with it, of the disdain towards science, order, and discipline.

> One cannot engage in science in general, and in physics in particular, assuming as a postulate that the universe is, as they stylishly say, a *projection of our own spirit.* [...]. Science denies idealism, and that is why it retains an ethical value: because one cannot impose discipline, abnegation, and sacrifice on others without imposing them in the name of something which is external to us.[25]

Scientists have high ethical values precisely because they are scientists, Garbasso claimed. National sentiment is inherent in the "scientific spirit," and thus depreciating science is the same as depreciating the nation. This is what Croce, according to Garbasso, had been doing.

> When the war broke out, the Patriarch of idealistic philosophers, Benedetto Croce, advised young scholars to shy away and prepare their academic qualifications so to have them ready for when everything was over. The World War was not, evidently, a projection of his spirit, and so it did not exist; he spent those suffering years more profitably, translating into Italian – more or less – Goethe's poems. But there is no need to recall such miseries to persuade ourselves that only realistic philosophy (which does not mean materialistic or positivistic) represents our true national philosophy.[26]

Croce's idealism, according to Garbasso, by overlooking science opened a fracture within Italian culture – a fracture that, needless to say, fascism had welded. The same idea was expressed by Agostino Gemelli, who unwaveringly declared that Idealism "has been and still is harmful to Italian science," having

24 Antonio Garbasso, "Il contributo degli Italiani alla Fisica," in *L'Italia e la scienza*, cit., 69.
25 Garbasso, "Il contributo degli Italiani alla Fisica," cit., 69.
26 Garbasso, "Il contributo degli Italiani alla Fisica," cit., 69.

"disturbed and eventually broken up" the unity of science and philosophy which characterized Italian thought at least since the Middle Ages.[27] This circumstance had become even more worrying since Idealism had entered high schools thanks to Gentile's reform:

> In the country of Cesalpino, Redi, Morgagni, Spallanzani, Volta, Marconi, Galvani, Schiaparelli, and Pacinotti, young philosophers in high schools look with disdain at scientific pseudo-concepts, happy as they are to have in the palm of their hand the concrete concept, and to have reduced nature itself to an act of thinking.[28]

Gemelli's criticism was widely shared by scientists, as they thought that Gentile's reform was devaluating scientific education. Two provisions, in particular, concerned them the most: the suppression of the physics and mathematics section in technical schools, and the creation of the *Liceo scientifico*, a high school with a limited science curriculum and somehow inferior to the humanistic-based *Liceo classico*. This meant that students coming from technical schools were prevented from enrolling in a University and those from the *Liceo scientifico* were allowed only a limited set of course choices, the *Liceo classico* being intended as the ruling class school and thus the only one granting access to all University faculties. In 1923, the *Accademia dei Lincei*, presided over by famed mathematician and antifascist Vito Volterra, issued a document that highlighted how the imbalance between philosophy and science in high school education could led to "the resurgence of those aprioristic, verbose, and meaningless ways of reasoning against which the great Renaissance spirits engaged in strong fights that we thought were definitively over after Galileo's victory."[29] This view was pretty much shared by the *Mathesis* society (presided over by Enriques), and by fascists and antifascists alike. Garbasso himself wrote to Volterra explaining that in Florence "everyone is dismayed" and wondered whether Mussolini could be somehow "enlightened."[30] What Gemelli, Volterra, Garbasso, and others feared was that science was being removed from the Italian cultural panorama, and that this would eventually undermine scientific research and, in the end, society's well-being.

---

27 Agostino Gemelli, "I rapporti di Scienza e filosofia nella storia del pensiero italiano," in *L'Italia e la scienza*, cit., 389.

28 Gemelli, "I rapporti di Scienza e filosofia," cit., 389.

29 Quoted in Giorgio Israel, "Vito Volterra e la riforma scolastica Gentile," *Bollettino dell'Unione Matematica Italiana*, serie 8, n. 3 (1998), pp. 269–287: 285.

30 Quoted in Lucio Russo, Emanuela Santoni, *Ingegni minuti. Una storia della scienza in Italia* (Milano: Feltrinelli, 2010), p.

So, another main topic thoroughly discussed at the conference was the centralization of scientific research. As mathematician Ettore Bortolotti pointed out, Italy had always been able to deploy an imposing front of "moral forces" in every field of knowledge, including science; yet, due to lack of organization, its results had often been overlooked by Italians or usurped by foreigners.[31] Italy had theretofore been able to keep up with more advanced countries thanks to individual geniuses who elbowed their way through adverse political conditions, but now things had changed. As specialization and technological advancements required more and more organized and well-funded research, the Italian genius had to give way to the better organization of foreigners. "Science has progressed to such an extent that in well-organized laboratories scientists can profitably work even if they are no geniuses," Garbasso added in his speech.[32]

In this perspective, the priority claims laid by historians of science should be aimed at mending the threads of the Italian scientific tradition, so to serve for a better organization of research. As Giulio Provenzal argued,

> This review will be accepted by everyone the day Italian chemists earnestly take into consideration the history of chemistry, whose main aim is to restore, in the Homeland's interest, the historical heritage of Chemistry itself, but also – and this is unperceived by many – to speed up the pace of scientific discoveries by looking for their fruitful roots in the hitherto unpublished and forgotten manuscripts and books of the most excellent Italian scientist, as they still bode new successes. It is pointless trying to lay priority claims on inventions foreigners have already acquired authorship on; it would be better to secure the discoveries and the brilliant insights still waiting to be studied, and thought over by Italians.[33]

The myth of Italian genius, the claim to Italian scientific priorities, the idea of a national tradition made of realism and experimentalism, centralization of scientific research: these were the ideological features that defined the image of the Exhibition in front of the fascist regime, the scholars' community, and the public itself. It is telling that Corsini had no part either in the press campaign or in the conference. Steering away from propaganda, he was at the time taking pain to preserve the fundamental nucleus of his original project to make

31 Ettore Bortolotti, "L'Algebra in Italia da Leonardo Pisano a Paolo Ruffini," in *L'Italia e la scienza*, cit., 53.

32 Garbasso, "Il contributo degli Italiani alla Fisica," cit., 96.

33 Giulio Provenzal, "Pagine di storia della chimica," in *L'Italia e la scienza*, cit., 114.

it outlive the ephemeral infatuation of the regime. In an interview given to the newspaper *la Nazione* on May 12, 1929, Corsini stated that the main aim of the Exhibition was to "make people understand how important it is to create places where these materials [*scientific collections*] can be appropriately kept [...] and to establish an Institute and Museum of History of Science furnished with a large library and a bibliographic office for the benefit of scholars to improve history of science studies as a whole."[34]

3 History of Science and the Public: Visitors at the Exhibition

The ideological profile of the Exhibition conflicted to some extent with the deeply-rooted Italian localism deplored by Mieli and made apparent by the set-up of the Exhibition itself. The adoption of a geographical ordering principle, decided as it was to comply with the conditions established by the local committees, prevented organizers from showing the "threads of the Italian scientific tradition" and eventuated in the Exhibition being, more than an exposition of history of science, an exposition of Italian scientists' relics. Once put aside any attempt at chronological or disciplinary consistency, what remained were, in fact, individuals; and furthermore, as the organizers had granted local committees the power to manage their own exhibits, the same individuals were linked to each other by the mere fact that they all happened to be Italians. So, the Exhibition had the overall appearance of a "topography" of Italian science, with each city celebrating its own scientists in its own fashion.

The geographical arrangement posed serious challenges to the visitor, as art historian Ugo Ojetti, who visited the Exhibition on October 8, 1929, eloquently pointed out. Ojetti was one the most influential personalities of the Italian cultural scene; he had served as director of the newspaper *Corriere della Sera*, and currently was the editor of the Art section of the *Enciclopedia Italiana*; moreover, he had already curated important art exhibitions – such as the "Exhibition of the Italian Portrait Art from 1500 to 1861" (1911) and the "Exhibition of the XVII and XVIII century Italian Painting" (1922) – and would later curate a huge exhibition on Giotto, the *Mostra Giottesca* of 1937.[35] In a way, Ojetti's opinion about the First Exhibition of History of Science is emblematic of the art historians' approach to historical scientific collections in general, and broadly

34 "La mostra di storia della scienza e gli alti compiti che essa si prefigge. Nostra intervista col vice-direttore Andrea Corsini," *la Nazione*, May 12, 1929.

35 See the entry "Ojetti, Ugo" edited by Laura Cerasi in DBI for a full bio.

explains why people like Mieli and Corsini received little support from them. According to Ojetti, even though the aim of the Exhibition was commendable, it could be hardly achieved, as in science, he argued, "it is the whole chain that matters; and as for the individual links, they matter only because they are held together in that chain."

> What impressed me the most was the number of things that I didn't know, and which tomorrow I won't know anymore. Distinguished professors, full of insignias, medals, bows, and diplomas, engraved in portraits, recalled in printed biographies, celebrated in sonnets: I was seeing their face and even hearing their name for the first time in my life. Physicists – or rather, protophysicists – chemists, astronomers, zoologists, botanists, physiologists, physicians, surgeons. After all, they are the men to whom we owe everything, our health, our comfortable life, [...] and yet they have been forgotten: a bust at the Pincio and a name for a street, maybe, but nothing more. That is unfair, isn't it? There are mediocre artists whose works are on display at the Uffizi Gallery or at Brera so that anyone can admire them; they are kept dusted, clean, complete with their tag and the artist's full biography; hundreds of people are willing to spend considerable sums of money just to own one of them, and will rejoice if they achieve in doing that. Also, there are mediocre writers whose books you had to read and sometimes even learn by heart in school: isn't that a contradiction?
>
> Well, I do apologise to the diligent organisers of this Exhibition, but it is precisely while seeing the discoveries of science and the discoverers themselves so well lined up, like the links of a chain, that I have realized that ignorance and oblivion are ineluctable, and that we, the public, cannot be blamed for that. I mean, yesterday's science matters because it created the science of today; and today's science, only because will create the science of tomorrow. In a word, it is the chain that matters; and as for the individual links, they matter only because they are held together in the chain. If the chain is broken up, the links themselves are nothing but scrap metal.[36]

At the end of his four-hours visit, Ojetti felt like "an old encyclopedia disarranged by the wind." The Exhibition, he explained, demonstrated that, contrary

36 Ugo Ojetti, "Le scienze in mostra," in *Cose viste. 1928–1930, tomo quinto*, (Milano: A. Mondadori, 1937), 132–141: 133–135.

to Corsini's opinion, historical scientific collections could not be equated with artistic and literary collections. In the latter fields "the opus is connected with its author just like light is connected to the Sun." In science, on the other hand, the opus, being "impersonal and unrecognizable," is detached from its author. A 'true' exposition of history of science, according to Ojetti, should have exhibited "the whole chain," not the disarranged individual "links."

The journalist and writer Corrado Sofia had a slightly different opinion. For him, the Exhibition should be considered as an initial index of the Italian historical scientific collections, the first step towards the establishment of a permanent museum, not as an exact representation of the Italian scientific development. Seen from this perspective, Sofia argued, the Exhibition's flaws could be easily excused.

> It does not matter if a tag attached on a device read (and still reads) "date of creation: 1600," but if you turn the device over you read "1550." It does not matter that nothing belonging to Schiaparelli and Stoppani is on display, that the Exhibition is not arranged by arguments, that the number of the books on display is far too large if compared to that of instruments and machines, that freshly varnished instruments are too shiny, that the very first typewriter (which was invented by an Italian) is nowhere to be found, that there are mysterious devices and gears without any explanation of their functioning: [...] all this does not matter, I say, because you can immediately notice that the Exhibition has been organised with great enthusiasm and passion. [...] I think that it is nothing but an admirable attempt, and yet, even from this first attempt you can clearly gather how beautiful and useful a permanent exposition of history of science could be.[37]

Mieli's collaborator Giuseppe Montalenti, who had also curated the set-up of the Rome room, expanded on these topics in a review published on *Archeion*.[38] The Exhibition, Montalenti argued, was historically flawed and had little didactic appeal; yet, it was able to speak to the visitor's imagination, reminding him of being the heir to people like Galileo, Redi, Malpighi, Volta, and so on:

> Let us serenely consider the results of the Exhibition and its aims. Firstly, the Exhibition has actually been staged, the old instruments and manuscripts have been dusted and cleaned up, and everybody can see them. Can you really dismiss that as "nothing"? Reminding the public that we

37 Corrado Sofia, "Profano alle prese con la scienza," *il Tevere*, May 29, 1929.

38 *Archeion* was the new name of the *Archivio di storia della scienza*.

> are the owners of all this immense patrimony of relics and testimonies of our glories and having everyone look at this beautiful collection of objects, making them live again part of their past life without drawing upon the libraries' heavy-going books! Recalling that we are the nephews of Leonardo, Galileo, Malpighi, Volta...![39]

The Exhibition's ability to excite the curiosity of the visitor is confirmed by several reports written by writers and journalists who had no scientific background. Journalist Antonio Baldini, a.k.a. *Melafumo*, wrote that, while visiting the Exhibition, he had felt "a compelling curiosity growing inside" him, even though he was "substantially unable to understand science."[40] Another journalist, who signed himself simply as G.P., reported that the Exhibition was, above all, "an enjoyable collection of countless curious rarities."[41] An account written by Florentine writer Raffaello Franchi conveys this kind of impressions more distinctly:

> So we went [to the Exhibition], and, much to my shame, I cannot tell you how much my companions enjoyed it. Save for the show of a horrified painter running away in front of vivisections and anatomical waxes revealing the deformities caused (may God help us) by some maladies [...] we haven't had any exciting adventure. Still, an air of pathos was gusting all around us and we were sure of feeling certain thrills and impression, such as when we entered the room of lighthouses [...]. In this small room, the human being, forgetting the smallness of his body as compared to the size of the world, feels like physically being the master of the universe.[42]

And one *Lallo* reports, in rather grandiloquent terms:

> Visitors – and here you may happen to hear dialects and languages from all around the world – enter the rooms with religious curiosity and observe the old manuscripts and the ancient machines; they follow the scientific advancements of a community, a region, a city, with the same intense emotion that can be felt when an illustrious and prestigious orator illustrates in front of an attentive crowd a masterpiece of art. This is the small desk, the humble table, patiently quilted by woodworms,

39 Giuseppe Montalenti, "La Prima Esposizione Nazionale di Storia della Scienza a Firenze," *Archeion*, 11 (1929), 239–241: 240.

40 Melafumo, "La Mostra della scienza ovvero Giannettino lasciato a se stesso," *la Tribuna*, May 19, 1929.

41 G.P., "Alla mostra storica delle scienze," *il Corriere Padano*, June 26, 1929.

42 Raffaello Franchi, "Oriente Fiorentino: Meriano," *l'Italia letteraria*, October 20, 1929.

> whence the first notions of a science that would revolutionize the entire field of Mechanics were taught a century ago; and the heart is caught in a grip: they are small desks, but how many lights they radiated throughout the world![43]

Without doubts, to create that sense of "religious curiosity" contributed, more than the "intense emotion" experienced while following the scientific advancements of a community, the difficulty to fully understand the scientifical and historical significance of the objects on display. "Once you get inside," Sofia noticed, "you turn to the left and to the right, and you find yourself alone, without a catalogue to guide you, at the mercy of tags and showcases, looking at imperfectly disposed objects."[44] The layman visitor would often find himself in front of mysterious devices, unknown machines, indecipherable manuscripts, books filled with obscure diagrams and figures, austere portraits of unheard-of people, and his first reaction was to try to enjoy them as art objects. And after all, Ojetti noted, being appreciated as art objects was the only way the objects on display could be rescued from the oblivion of time.

> I have seen a glare of that fear [i.e., of the oblivion of time] in the elegance with which even the sternest scientists adorned their most precise instruments and their most pondered writings, almost as if they were appealing to Art for assistance in traversing the river of time [...]. Compared to those prettiness, today's shiny and unembellished instruments are like sharp blades. The Astronomer, the chemist, the physicist, the biologist: he is afraid of everything unnecessary in his struggle against mystery. Yet, if he wants to get in touch with the world and catch our attention as public, he has to turn to that charm, that common language that made him human among human beings.[45]

Organizers were fully aware of the visitors' troubles and tried to find a solution to them. Guided tours, held by historian Pietro Pagnini twice a week, were launched on May 23;[46] brief descriptions of the main rooms were published in newspapers; and an Official Guide written by Umberto Repetti was published in August by the EAT. It could be purchased at the Palace of Expositions and

43 Lallo, "Una visita alla mostra della scienza," *Terme e riviere*, October 29, 1929.
44 Sofia, "Profano alle prese con la scienza," cit.
45 Ojetti, "Le scienze in mostra," cit., 139–140.
46 "Una guida pratica per le visite alla Mostra della Scienza," *la Nazione*, May 22, 1929.

allowed visitors to get at least a general idea of what was on display. These actions, however, though undoubtedly helpful to the visitors, proved less effective than expected, undermined as they were by the Exhibition's geographical ordering. To many, the explanations provided by Pagnini and those written out in the Guide looked like a disorganized cornucopia of information – a cornucopia which even learned men like Ojetti found it hard to manage.

> Twice a week, at fixed hours, a diligent professor guides visitor through the Exhibition, explaining everything from the forceps up to the airplane, and I have even tried to follow him, together with college students, draftees, seminarists, and modest young ladies accompanied by their mothers and intent upon writing things down in their notebooks. But the worst, here, is that the more you learn, the more the sea of your ignorance looks dazzling and endless, the rainbow of progress painted on that sea by science itself being of no solace.[47]

Such issues notwithstanding, the Exhibition attracted a number of visitors well beyond the organizers' expectations. The success was mostly, though not exclusively, due to the support given by the fascist regime, which admonished that paying a visit to the event was "a duty every true Italian owes to his country."[48] Moreover, the Government established reductions on train fares, while the Executive Committee encouraged group visits by selling tickets at sale price. According to official estimates, about 100.000 people visited the Exhibition over the six-months period of its opening, with a peak on Sunday the 25th of August, when there were about 10.000 visitors.[49] Though far from the numbers routinely achieved in occasion of art exhibitions – not to mention Universal Expositions – it was still a great result for an unprecedented and supposedly unappealing event. Unsurprisingly enough, such a success was greatly celebrated by fascist propaganda as a clear evidence of "the high degree of moral evaluation and civil dignity whereto Fascism has brought the masses, which were once humiliated, dehumanized, contaminated with beastly doctrines."[50]

---

47 Ojetti, "Le scienze in mostra," cit., 140.

48 "Alla mostra delle scienze," *Nuovo Giornale*, October 27, 1929.

49 "Diecimila persone alla mostra della scienza nella giornata di domenica," *la Nazione*, August 27, 1929.

50 "60.000 visitatori in 14 giorni alla Mostra della Scienza," *Unità Cattolica*, August 28, 1929.

4 History of Science and Entertainment: Music, Cinema, and Entertainment at the Exhibition

Alongside the regime's support, the City of Florence's initiatives, and the Executive Committee's promotions, a fair share of the Exhibition's success can be credited to the so-called "Evening receptions" held at the *Parterre*. They were music concerts, projections, and other recreational events that livened up the Florentines' Summer evenings. As the news correspondent for *il Giornale d'Italia* wrote on August 30,

> I have often seen youngsters and parties leaving the city center and heading to the San Gallo *Parterre*; I have also very often seen entire families heading to the *Viali* and spend the evening in the wonderful garden of Piazza Cavour. In fact, according to the statistics I have been allowed to peruse, the number of visitors of the Exhibition has significantly increased since when the Executive Committee decided to hold evening receptions. Lately, the number of people who visited the Palace of Expositions has truly been extraordinary.[51]

The idea of evening receptions came from the opera-enthusiast Ginori Conti and was in many respects an effect of the emphasis put by the regime on cinema, radio, and theatre as means of propaganda. The opening event was a series of two solo concerts by young pianist and composer Marcello Boasso, held in the Palace of Expositions' conference room on June 19 and June 23. Born in Turin in 1902, Boasso, who had been a student of Ferruccio Busoni, was one of the most respected pianists and conductors of the time. The Florence concerts were part of worldwide tour titled *L'Italianissima* he had embarked on in 1928. On both dates, Boasso played to a packed house pieces by Bach, Mozart, Schubert, Bellini, Musorgskij, Rimsky-Korsakov, and of his own composition.[52] The concerts were such a hit that the Executive Committee decided to organize another one for July 14. On that occasion, thanks to an amplification system devised by Raffaello Bacci, the concert was transmitted live to a public of about 2500 people gathered outside the Palace of Expositions.[53]

Meanwhile, on July 2, summer opening times came into force. The Exhibition, until then open to the public from 9 A.M. to 12 A.M. and from 4 P.M. to 8 P.M., was now open from 9 A.M. to 12.30 A.M. and from 6 P.M. to 11 P.M.

51 "Le audizioni alla Mostra della scienza," *il Giornale d'Italia*, August 30, 1929.

52 "Due concerti di Marcello Boasso alla Mostra della Scienza," *la Nazione*, June 15, 1929.

53 "I concerti all'aperto col microfono," *La propaganda musicale*, II, September 1929.

The garden was adorned with festoons and decorative lighting; an outdoor coffee-house was set up so that visitors could sit and listen to live music or to what the Executive Committee called, not without pomposity, "gramophonic performances;"[54] and, on a huge screen placed in the northwest corner of the garden, newsreels about the regime's pretended military power and Italy's supposed technological advancements ran constantly.[55] These initiatives were aimed at getting people, especially those who usually showed no interest in science, go visit the Exhibition. Communicating history of science to a broad public audience and gaining its support was vital to the organizers' program, but, as a columnist wrote, "the public deserves to be helped."[56]

Still, not everyone agreed with the strategy of blending culture and entertainment together. *Il Bargello*, official journal of the Fascist Provincial Federation of Florence, for example, harshly criticized it:

> An opera season is currently being held at the San Gallo *Parterre*, and it suffices to look at the exquisite playbills marked by extraordinary typographical bad taste posted by the E.A.T. everywhere in the city to figure out how astounding it is. We are talking of a collection of crummy "gramophonic performances," announced with all the pomp of a kitschy provincialism. They announce, for example, Giacomo Puccini's *Boeheme* – an opera in four acts etc.–its interpreters and conductors, just like as if it was an actual performance, and so on for all the other records. Besides, the *Parterre*'s garden is all adorned with Japanese lighting – what a wonderful effect! Truly a festival. Then, inside the Palace of Expositions, there is the Exhibition of History of Science, a very serious thing. Now, we don't understand why they should be serious on the inside and ridiculous on the outside![57]

The aforementioned *Lallo* – the same journalist who wrote about the visitors' "religious curiosity" – attacked almost the same way. "Florentines from Florence ... and from Milan, are right: that fun fair of games, that permanent carnival of nursemaids and soldiers – and I cannot figure out why they put it there – obstructs and annoys."[58]

54 "Le audizioni all'Esposizione di storia della scienza," *Unità Cattolica*, August 10, 1929.
55 "Rappresentazioni cinematografiche alla Mostra della Scienza," *la Nazione*, July 24, 1929.
56 "Le audizioni alla Mostra della scienza," cit.
57 "La scienza e il festival," *il Bargello*, September 15, 1929.
58 Lallo, "Una visita alla mostra della scienza," cit.

This kind of criticism shows on the one hand the persistence of an aristocratic idea of culture even among local newspapers, while indirectly confirming, on the other, that the Exhibition's organizers had reached their purpose of bringing history of science to the public. The concerts, the outdoor coffee-house, and all the other initiatives not strictly related to history of science recreated on a lesser scale the joyful mood of Universal Expositions, decisively contributing to the success of the event: it was, as Franchi put it, "like being invited to party in the holy enclosure of ancient and modern discoveries."[59]

59 Franchi, "Oriente fiorentino," cit.

CHAPTER 5

The National Museum of History of Science

1 From the Exhibition to the National Museum

The First National Exhibition of History of Science gathered together almost 10.000 items coming from about 80 cities and 200 public and private lenders. A large part of these items was on public display for the first time ever, while others had been discovered in warehouses thanks to an unprecedented research activity which had involved librarians, archivists, historians, and museum curators all over Italy. Albeit within the limits of its own set-up, the Exhibition succeeded in bringing history of science to the public's attention, and significantly contributed to put the issue of the protection of historical scientific collections on the political agenda of the Italian state. Thanks to this success, the debate about the necessity of establishing a National Museum of History of Science, theretofore limited within a narrow circle of scholars, made its way to the general press, reaching a wider audience.

Since the very opening of the Exhibition, several independent articles (that is, articles not issued by the Exhibition's press office) had been signaling the urgency of protecting and promoting historical scientific collections, putting forth the idea of creating one or more special museums to take care of the objects which were on display in Florence. On May 18, 1929, writer and journalist Italo Del Giudice, who would later become a successful science popularizer, pointed out "the benefits of creating permanent collections and museums to preserve the Exhibition's materials;"[1] in July, Mario Giulia called on the Government to "turn the Exhibition into a permanent Museum of History of Science;"[2] in October, Giulio Provenzal expressed his wish that such a museum retained, "in original or in copy, as many evidences as possible of the Italian scientific activity;"[3] and the National Association of the Fascist Teachers explicitly asked that the new museum, in the event, be set up in Florence:

1 Italo Del Giudice, "Visitando la Mostra nazionale di Storia delle Scienze," *il Corriere d'Italia*, May 18, 1929.
2 Mario Giulia, "La Mostra fiorentina di storia delle scienze," *l'Illustrato*, July 21, 1929.
3 Giulio Provenzal, "Alla Mostra di Storia della Scienza a Firenze. La scienza italiana rivendicata," *Scienza e Vita*, I, 2 (October 1929), 198.

 | DOI:10.1163/9789004297418_006

> Since Florence has such a valuable historical material both in terms of quality and quantity to establish by itself an *Institute of History of Science*, [...] we can't help but fervidly asking that in Florence, which enshrines our most beautiful scientific glories summed up in the divine names of Leonardo and Galileo (not to mention countless others), be established that Institute of History of Science, which already exists in embryo, and to the constitution of which would suffice one simple signal from the Duce.[4]

Public stances of this kind show how the Exhibition succeeded both in pointing out the importance of historical scientific collections as 'cultural objects' and in conveying the image of Florence as the cradle of Italian science. The Florence exhibit, staged in the Palace of Expositions' largest room, centered on Galileo and the Galilean tradition, and surrounded by smaller rooms housing the other cities' exhibits, physically substantiated the claim that Florence itself was the very cradle whence Italian science had risen. By the Exhibition closing, organizers felt confident that the project of establishing a National Museum in Florence had earned the support of the Italian cultural environment, and that the same environment had also mostly accepted the identification of the Galilean tradition with the Italian scientific tradition as a whole. Moreover, the main competitor for that project, namely, the National Institute of History of Science in Rome, had *de facto* dissolved in the meanwhile, as Mieli had left Italy for France in 1928, and Enriques, absorbed as he was in his tasks as professor at the University of Rome, president of the *Mathesis* society, editor of the *Periodico di matematiche*, and director of the mathematics section of the *Enciclopedia Italiana*, was unable to attend to it. And so, with the 'Roman threat' warded off and the Exhibition showing the richness of the Florentine collections, the organizers' confidence grew. Even if none of the local committees had ever taken into consideration the possibility to leave its collections in Florence, Corsini, Ginori Conti and Garbasso felt that they could still set up a National Museum. "To create a National Museum here in Florence," Ginori Conti highlighted in the speech delivered during the closing ceremony of the Exhibition, "we don't need to deprive universities' scientific cabinets and civic museums of their collections. What already is in Florence is more than enough for us to create a one-of-a-kind Museum."[5] The National Museum, according to its Florentine promoters, would bear witness of the Florentine roots of

4 Dom., "Glorie italiche alla grande Esposizione di Storia delle Scienze," *Scuola fascista*, October 27, 1929.

5 "La chiusura della Mostra Nazionale di Storia della Scienza," *la Nazione*, November 12, 1929.

Italian science and, along with the *Uffizi* Gallery and the National Library, give substance to Florence's primacy in science, literature, and art.

But the organizers felt confident also because of another, more prosy reason. The Exhibition in fact turned out to be an economical success, having closed its balance with a profit of about 80.000£. According to the balance sheet of December, 1930, the Exhibition registered a revenue of about 970.000£ against charged expenses of about 890.000£; from ticket sales alone, it had raised 163.000£, more than five times than expected.[6] This success was especially important because, as the propagandistic potential of the Exhibition exhausted itself, the regime's interest in the Florentines' initiatives began to fade. In fact, though acknowledging the importance of establishing a National Museum of History of Science, the Government was unwilling to invest any money in it, preferring to direct its attentions to the CNR instead.

2 Administration and Financing

The National Museum of History of Science created in the wake of the Exhibition of History of Science was an unusual kind of institution within the Italian museum constellation, since it was to be administered together by the Government, the City of Florence, and the University. Nominally, it was a National Museum, having been granted that status by Mussolini himself and being meant to serve 'the whole nation'; yet it was kept alive not by the Italian Government but by two local institutions, just like any other Italian civic museum was; finally, the largest part of its collections came from the University of Florence.

From the administrative point of view, the 'new' institution stood on the Institute of History of Science, having inherited the latter's charter and organizational structure. The Museum was indeed to be managed by an executive board consisting of representatives from the University of Florence, the Ministry of Public Instruction, the City of Florence, and the Private Donors. The board would elect a president, who, in turn, would appoint a director. The first board consisted of Giorgio Abetti, Giotto Dainelli, Antonio Garbasso, Piero Ginori Conti, and Luigi Rolla. Predictably enough, Ginori Conti was elected as president, and Andrea Corsini was appointed as director. Though this new 'National Museum' managed its own funds, appointed employees, and negotiated purchases, it still was the same old Institute of History of Science and

6 "Conto Consuntivo – Dicembre 1930," ASCFi, CF9260, s. 7.5, 371–425.

continued to depend from the University. In fact, according to the agreement signed with the University of Florence in 1933,

> 1. The R[oyal] University of Florence consent that the Institute of the History of Science should be entrusted with all of the *Museo degli strumenti antichi di Fisica e Astronomia*'s collections, and with the instruments, objects and collections of historic value currently held in the University's institutes [...]. The University also consent that the said Institute should be entrusted with all the books and works of historic value forming the *Museo degli strumenti antichi di Fisica e Astronomia*'s library that is presently stored in the Physics Laboratory.
> 2. [The University of Florence] consent that all the said materials, along with those already owned by the Institute of History of Science and those that will be acquired in the future, should be known under the general denomination of "the University of Florence National Museum of History of Science."

It being understood that the museum's operating costs and all the expenses necessary for the maintenance of the collections should be covered by the Museum, the collections themselves remained in the "absolute possession" of the University, which would decide about loan policies and restoration projects. Every object on display should bear a tag signalling that it was property of the University of Florence, and a clearly visible sign, reading, "University of Florence – National Museum of History of Science," should be affixed to the Museum's entrance. The University also requested that the Museum be open to the public at least three days a week; the Museum, for its part, would be granted any proceeds generated by ticket sales.[7]

The main tasks the Museum's administrators had to focus on, at this early stage, were pointed out by Ginori Conti during the first meeting of the executive board held on December 6, 1930. It was necessary, according to Ginori Conti, to open the museum to the public as soon as possible; a provisory index of the collections and of the library had to be compiled; and an effective acquisition policy should be developed. Corsini, speaking in his capacity as director, underlined that the museum needed to be regularly subsidized, as at least a full-time janitor, a curator, and a librarian had to be appointed in the first place. So, after a brief debate, the board agreed to file requests for financial support at the Italian Government, the Provincial Council for Economy, and the City of Florence; an association called "the Museum's friends" was created with the

7 "Convenzione tra l'Università di Firenze e l'Istituto di Storia della Scienza," ArchMG, IMSS, 1933.

purpose of gathering privates' donations; and Garbasso offered to uphold the cause of the museum within the newly-established *Accademia d'Italia* ("the Italian Academy").[8]

Thus, on December 15, Ginori Conti wrote to Mussolini. In the letter, the Prince briefly outlined the history of the museum, remarking the success of the First National Exhibition of History of Science and the fact that the museum had not theretofore weighed on the nation's coffers, and asked, on behalf of the museum's executive board, for a one-time contribution of 100.000£ and for an annual subsidy of 25.000£. "The work we have done so far must not be lost," Ginori Conti wrote, "on the contrary, it is necessary to continually enhance the museum."[9] On January 13, 1931, Mussolini wrote back, explaining that, even though the Government fully acknowledged the museum's "high aspirations," it was impossible, due to the current financial situation, to fulfil the Executive Board's request.[10] The Great Depression was hitting hard: just a few weeks before replying to Ginori Conti, Mussolini had issued a circular letter addressed to the Ministries urging them to come to terms with the "difficult circumstances" and avoid unnecessary outlay.[11] The largest share of the limited funds available for cultural and scientific purposes were to be allocated to the CNR or to initiatives more valuable in terms of propaganda, like the Exhibition of the Fascist Revolution (*Mostra della Rivoluzione Fascista*), which was scheduled for 1932 and was meant to celebrate the tenth anniversary of the March on Rome.[12]

However, the executive board had better luck with the requests sent to the City of Florence and to the Provincial Council for Economy, as they approved annual contributions of 8000£ and 4000£, respectively. The Italian Academy, even though unable to allocate annual contributions, promised non-recurring allowances in the amount of up to 3000£ as "incentive bonuses" for the Museum.[13] Thus, when the executive board met on March 31, 1931, the Museum's financial situation, albeit not quite rosy, looked comforting. Starting from then on, the institution would receive annual contributions for 12.000£, enough to appoint a janitor, a curator (Giulio Cipriani), and a librarian (Rosa

8 Minutes of the meeting of December 6, 1930, ASCFi, CF9260, s. 8, 427–428.

9 Ginori Conti to Mussolini, December 15, 1930, ASCFi, CF9260, s. 1, 195r–198r.

10 Mussolini to Ginori Conti, January 13, 1931, ArchMG, IMSS, 1930–1931.

11 Document reproduced in "Spending review anni Trenta," *Gnosis. Rivista Italiana di Intelligence*, 2 (2012).

12 On the Exhibition of the Fascist Revolution see, Antonella Russo, *Il fascismo in mostra* (Roma: Editori Riuniti, 1999), 33–73. The Exhibition index has been reprinted in 1990: *Partito Nazionale Fascista. Mostra della Rivoluzione Fascista. Inventario*, edited by Gigliola Fioravanti (Roma: Ministero per i Beni Culturali e Ambientali, 1990).

13 Ginori Conti to Corsini, March 13, 1931, ArchMG, IMSS, 1930–1931; The Italian Academy to Ginori Conti, April 28, 1931, ArchMG, IMSS, 1930–1931.

Franceschini); to cover the operating costs and to put a little cash aside. In December 1932, the Museum had about 6000£ deposited in a bank account.[14]

These resources, however, even if they allowed the Museum to be open to the public, were not enough to sustain the executive board's ambition of enhancing and enlarging it. Ginori Conti and Corsini were confident that, somewhere in the near future, Mussolini would change his mind and decide to give them financial support, but despite several request sent to the Government no support ever came and the Museum would survive for years to come solely thanks to its 12.000£ income, Ginori Conti's generous donations, and Corsini's passion and perseverance.

3 Location and Collections

As seen in chapter 2, in 1926 the Evaluation Commission appointed by the University of Florence to judge about the Institute of History of Science's chances of development had advanced the proposal to relocate the institute at the *Tempio degli Scolari*. The building, at that time, was nothing more than a ruin, but the Commission was confident that, thanks to Garbasso's intercession, the City of Florence would renovate it. Still, as the estimated costs of the renovation turned out to be too high, by 1929 Giuseppe Della Gherardesca, Garbasso's successor to the office of *Podestà*, had planned to hand the building over to the National Association for War Wounded and Disabled People (*Associazione Nazionale Mutilati e Invalidi di Guerra*), which was about to purchase part of the adjacent Santa Maria degli Angeli Monastery.[15]

So, the Executive Committee settled for Palazzo Castellani, a four-storey historic building located near the Uffizi gallery and built by the Castellani family after the Arno's overflow of 1333. Bought by Grand Duke Cosimo I in 1558, it had first served as the seat of the *Podestà*, and then, from 1573, of the *Giudici di ruota* (Judges of Civil Cases). After the Italian Unification, the building became public property and was destined to temporarily house the Crusca Academy on the second floor and the manuscript collections of the National Library on both the ground and the first floor. In 1929, as the manuscript collections had been already moved to the National Library's new building and the Deputation for the History of the Fatherland (*Deputazione di storia patria*) had settled on

14 Ginori Conti to Corsini, December 6, 1932, ArchMG, IMSS, 1932.

15 See Dino Savelli, *La rotonda del Brunelleschi: storia e documenti* (Firenze: Esuvia, 1992), 17–18.

the first floor, fifteen rooms on the ground floor were vacant and available.[16] In November, Corsini and Ginori Conti made a deal with the State Property Office, according to which the Executive Committee of the First National Exhibition of History of Science was granted Palazzo Castellani's ground floor on condition that renovation expenses be covered by the Executive Committee itself.[17] The Executive Committee invested about 28.000£ to update the electrical system, install a brand-new linoleum flooring, paint the walls, and buy new furniture. Once the work was finished, Corsini had the Institute of History of Science's library moved to Palazzo Castellani, while Garbasso made a selection of artefacts amongst the *Museo degli strumenti antichi*'s collection to be put on display. On April 14, 1930, the new museum was visited by the rector of the University of Florence, Bindo De Vecchi,[18] and on May 18 the National Museum of History of Science was officially inaugurated in the presence of Mussolini. According to Garbasso's rather obsequious report, Mussolini had a look at Galileo's manuscripts and asked Corsini about military surgeons in Napoleon's army.[19]

Ginori Conti had asked the University of Florence to borrow part of the *Museo degli Strumenti antichi*'s holdings on April 14, 1930, receiving an affirmative response the very same day.[20] Garbasso, director of the *Museo*, had made a selection and then, towards the end of April, the collections had been sent to Palazzo Castellani.[21] For the most part, they consisted of relics coming from Galileo's tribune, tools and equipment related to the Accademia del Cimento scientific activity, astronomical instruments once in use at *la Specola* observatory, and apparatuses belonged to the *Imperial Regio Museo di Fisica e Storia Naturale*'s old laboratories. All these items were being added to the Santa Maria Nuova General Hospital's surgical collection the Institute of History of Science had been entrusted with in 1926, and to some reproductions and photographs offered by local committees.

---

16 On Palazzo Castellani, see Federigo Fantozzi, *Nuova guida ovvero Descrizione storico-artistica critica della città e contorni di Firenze compilata da Federigo Fantozzi architetto* (Firenze: presso Giuseppe Ducci, 1847), 155–156; Mario Bucci and Raffaello Bencini, *Palazzi di Firenze* (Firenze: Vallecchi, 1971), 15.

17 Minutes of the meeting of December 6, 1930, ASCFi, CF9260, s. 8, 427–428.

18 Corsini to De Vecchi, April 14, 1930, ASUFi, a. 1930, f. 597, fasc. 126.

19 See Antonio Garbasso, "Al Museo di storia della scienza," *il Bargello*, II, 21 (1930).

20 Ginori Conti to De Vecchi, April 14, 1930; De Vecchi to Corsini, April 14, 1930, both in ASUFi, a. 1929, f. 588, fasc. 126.

21 De Vecchi to Garbasso, April 26, 1930, ASUFi, a. 1929, f. 588, fasc. 126.

It is difficult to tell how this material was arranged originally.[22] As said above, there were fifteen room on Palazzo Castellani's ground floor; as the Museum's library and offices had been allocated in the five smallest rooms, the available exhibition space consisted of ten relatively big rooms. According to Garbasso's report about Mussolini's visit, the objects were arranged by discipline and there were rooms devoted to Astronomy, Physics, Medicine, Chemistry, and Natural Sciences.[23] Another report, possibly written by Corsini soon thereafter, summarily mentions the existence of rooms devoted to Galileo, the Accademia del Cimento, Girolamo Segato's petrified human remains together with Giovanni Antonio Brambilla's surgical Armamentarium, and Peter Leopold's Chemistry Cabinet, roughly confirming Garbasso's description.[24] According to historian of medicine Luigi Castaldi, who wrote an article about the Museum appeared in *Archeion* in 1931, the exhibition space consisted of nine rooms "filled with extremely interesting items documenting the past."[25]

An article written by journalist Piero Domenichelli and appeared on December 31, 1932, in *il Popolo d'Italia* gives us a more detailed account. The Museum, Domenichelli wrote, consisted of ten rooms. Upon entering, visitors were led to the left, in a room housing astronomical, gnomonic, and geodetic instruments; Evangelista Torricelli's barometer; and a collection of instruments, belonged to Vincenzo Viviani, featuring astrolabes, compasses, armillary spheres, solar and stellar clocks. The second room was piously dedicated to Galileo: visitors could admire Galileo's telescopes, an objective lens he had grinded and polished, the geometric and military compass he had devised and built, a magnet he had once presented to Grand Duke Ferdinand II, the thermoscope he had designed, an astrolabe he used to use, and his middle finger enshrined in a small crystal urn. In the following room, thermometers, odometers, compasses, goniometers, glass vases, and ampoules belonged to the Cimento Academicians were on display. Peter Leopold's chemistry cabinet, Segato's anatomical preparations, and Brambilla's surgical instruments were housed in two separate rooms, completing the first part of the exhibition. At this point, visitors had to walk back toward the entrance and head for the east wing of the palace, where two other rooms were filled with pneumatic pumps, more solar and stellar clocks, and Leopoldo Nobili's instruments for the study of Thermodynamics and Electrochemistry.[26]

---

22 On the Museum's setups, see Marco Beretta's detailed study, "Andrea Corsini and the Creation of the Museum," cit., 11–26.

23 Garbasso, "Al museo di storia della scienza," cit.

24 The report is presently held in ArchMG, IMSS, 1930.

25 Luigi Castaldi, "Il Museo fiorentino di storia delle scienze," *Archeion* 13 (1931), 66–67.

26 Piero Domenichelli, "Arte e scienza italiche sulle orme di una gloria immortale," *il Popolo d'Italia*, December 31, 1932.

The majority of these objects came from the *Museo degli Strumenti antichi*, and dated back to a period ranging from the 16th to the 19th Century. Shortly after the Museum's opening, however, Corsini and Ginori Conti set to work to augment the collections. They directed their attention to the Copernican Museum, whose collections had been on display at the Exhibition and were now stored at the Lincean Academy, waiting for a proper seat. Relying on the fact that the Florentine Museum had been recognized the status of National Museum, on November 3, 1932, Ginori Conti sent a request letter to Mussolini asking for them.

> Some years ago, as the famed Observatory of the Roman College closed down and the Prehistoric Museum broadened out in the Jesuits' Palace of Rome, the so-called Copernican Museum lost its humble premise. This Museum is called Copernican, yet it holds almost nothing belonged to Copernicus himself; instead, its holdings are, because of their authors and history, directly linked to Galileo and the *Accademia del Cimento*'s relics which I had the honour to show to Your Excellency on the occasion of the Museum's inauguration.
>
> The Copernican Museum, after the said events, was temporarily stored at the Lincean Academy, and since it could not be housed there for long, some time ago Senator Garbasso, Senator Millosevich, His Excellency Parravano, and I planned to move it to Florence, receiving a positive feedback from Balbino Giuliano, who was then Minister of National Education.
>
> Since we do firmly believe that, in order to avoid dispersion and money-wasting, the best choice would be to accommodate the Copernican Museum at the Museum and Institute of History of Science in Florence, I am now pleading Your Excellency – on behalf of the said Museum's executive board and on my own – to consent to that plan, so that in a not too distant future we hopefully shall be able to show You all those relics – which represent the origin of the experimental method – gathered together in a single and proper location and serving as teaching and incitement for Generations to Come.[27]

It is not difficult to recognize, in Ginori Conti's words, the echo of the identification between the Galilean tradition with the Italian tradition. In fact, the Prince's point is that the Copernican Museum was improperly called "Copernican," since nothing or almost nothing were to be found, amongst its collections, that had actually belonged to Copernicus. On the contrary,

27 Ginori Conti to Mussolini, November 3, 1932, ArchMG, IMSS, 1932.

those collections could be "directly" linked to the Galilean tradition which was on display in Florence, and thus moving the Copernican Museum to Florence would signify putting it in its 'natural' environment, glorifying it to the full. Ginori Conti was very confident about the outcome of the plan. On December 6, he wrote to Corsini informing him that, together with the rector of the University of Florence, he had been studying the better accommodation for the Copernican Museum, coming to the conclusion that the best thing to do was to store it in Palazzo Strozzi until the second floor of Palazzo Castellani was available.[28] The plan, however, turned out in nothing, as the Copernican Museum would remain in Rome to be definitively quartered, together with the Astronomical Observatory, at Villa Melloni (1935).

The Museum, however, did not give up the ambition of becoming Italy's main conservation center for historic-scientific artifacts, relying on the uniqueness of the institution and on its highly patriotic mission. This ambition became increasingly apparent after 1933, following the Museum's involvement in the Chicago World Fair (1933–1934). The Exposition celebrated "a century of progress,"[29] and the Fascist regime considered the event as a valuable opportunity to present itself as the restorer of the Italian scientific glory. On that occasion, Guglielmo Marconi, who, as president of the CNR, was to coordinate the Italian participation, commissioned the Museum to produce four sets of replicas that were to be sent to the Chicago Exhibition, the Science Museum in Münich, the Museum of Science in London, and the CNR in Rome. The models built in 1932 for *A Century of Progress* under Cipriani's supervision featured instruments developed by Torricelli, Viviani, and Nobili, as well as Galileo's telescopes. After the exposition closed in 1934, some of them were donated to Chicago's Museum of Science and Industry.[30] As the Italian participation at the Chicago World Fair turned out to be a huge success, Ginori Conti took the occasion to write again to Mussolini to reaffirm the primacy of the National Museum of History of Science and underline its 'national' mission:

> The Museum – Ginori Conti wrote – has continued to improve and develop; still, to fully take on its national mission, it needs to be fully recognized by the Government, that is, it should be consigned all the

28 Ginori Conti to Corsini, December 6, 1932, ArchMG, IMSS, 1932.

29 For a general overview on the 1933 Chicago World Fair, see Lenox R. Lohr, *Fair Management. The Story of* A Century of Progress Exposition (Chicago: Cuneo Press, 1952); Cheryl R. Ganz, *The 1933 Chicago World's Fair: A Century of Progress* (Chicago: University of Illinois Press, 2012).

30 See Beretta, "Andrea Corsini and the Creation of the Museum," cit., 1–36.

> historical-scientific collections that at the moment are not properly quartered and copies of all the objects already on display in museums.[31]

The success of the Italian exhibits at Chicago, Ginori Conti went on, was also to be credited to all the work already done for the First National Exhibition of History of Science. Now, he wrote, the National Museum was expanding on that, striving to compile a general catalogue to "make Italians and foreigners know about our scientific glories." The same view was endorsed by Florentine historian of science Sebastiano Timpanaro, who, writing in the Milanese newspaper *l'Ambrosiano*, stated that the Museum should be considered "a central Museum" and thus be consigned copies of the most important scientific relics.

> Just to be clear, I do not expect Florence to seize all the national scientific material: every city is rightfully proud of and attached to its scientific glories. But the Florence museum could be consigned good copies, like the ones that has been sent to Chicago World Fair and are about to pass to the US Science Museum. [...] The central Museum of Florence shall not abolish scientific museums already existing elsewhere. It is good if there are many of such museums – and libraries, too – but that they must be places for studying, not depositories for dust.[32]

Even though the Museum failed to obtain the kind of institutional support both Ginori Conti and Timpanaro longed for, it was enlarged anyway. In 1934, as the National Library had begun to vacate its rooms on the first floor of Palazzo Castellani, the Museum could display its collections in 14 rooms spread over two floors. Exhibits were rearranged, and eventually a reading room and possibly a conference room were set up.[33]

4 Visitors

The Museum's early years were not easy. Enjoying no financial support from the Government, it relied solely on the limited budget provided by the City of Florence, the Provincial Council for Economy, the Accademia d'Italia's and Piero Ginori Conti's casual contributions, and ticket sales. Given these

31 Ginori Conti to Mussolini [n.d.], 1933, ArchMG, IMSS, 1933.

32 Sebastiano Timpanaro, "Illuminazioni scientifiche: Firenze per la storia della scienza," *l'Ambrosiano*, March 2, 1934.

33 See Beretta, "Andrea Corsini and the Creation of the Museum," cit., 19–20.

difficulties, though officially inaugurated in 1930, the Museum opened to the public only in April, 1931. In 1935, it was open every day from Monday through Sunday from 9.00 to 12.00 in the morning and from 14.00 to 17.00 in the afternoon; during summer, it was open from 9.00 to 12.00 and from 15.00 to 18.00. Tickets were being sold at the price of 2£.[34] The Museum Guestbook allows us to guess in broad terms something about the flow of visitors. Obviously, the data cannot be considered complete, as not all of the visitors necessarily signed the Guestbook; yet they furnish us with gross indications.

During the first five years of opening (from April, 1931, to April, 1936), the number of visitors who signed the Museum Guestbook amounted to 2292, with an average of 38 registered visitors per month and about 458 per year, which translates for just over one registered visitor per day. These numbers, being similar to those of more geographically peripheral civic museums like the Etruscan Museum of Cortona,[35] may look unimpressive, especially if compared to those of a typical National Museum.[36] Still, when considered within the context of the Florentine situation and if compared to those of the *Museo degli Strumenti antichi*, they appear differently. As Corsini had already remarked in 1924,[37] in Florence, the majority of visitors divided their time between the Uffizi Gallery and a few other art museums, with the result that a significant number of museums would often go unvisited: as Giovanni Poggi had stated in 1914 when he was summoned to report about the state of the Florentine galleries in front of the City Council, in Florence there were "too many museums."[38] The situation was especially bad for the *Museo degli strumenti antichi*, which was still suffering from the side effects of the dispersion of the *Regio Museo di Fisica e Storia Naturale*'s collections. In fact, being a little away from the main tourist routes, between 1922 and 1927 it had registered just 160 visitors or so, and that was one of the main reasons why since the early 20s Garbasso had been trying to have it moved somewhere else.[39]

With the opening of the National Museum of History of Science, the number of registered people visiting the *Museo degli strumenti antichi*'s collections in their new seat multiplied nearly fourteen times. An average 40% of those

34 Corsini to Grieben-Verlag, April 25, 1935, ArchMG, IMSS, 1935.

35 See *Polimnia* 7 (1930), 857.

36 According to the *Bollettino statistico mensile*, the *Museo Nazionale del Risorgimento* in Turin, for example, would routinely get about 10.000 visitors per year.

37 Corsini, "La Specola di Firenze," cit.

38 See *Atti del Consiglio Comunale di Firenze*, 1 (1914), 207–208.

39 See "MSA – Firme dei visitatori, 31 luglio 1875–6 ottobre 1937," ArchMG, MSA-Museo: inventari, [post 1864]-1954. On Garbasso's project, see ch. 2.

visitors were foreigners. Americans, Frenchmen, Germans, and also random visitors from South America, Japan, Hungary, and Czechoslovakia signalled an inherent international vocation that the Museum would maintain in the years to come as well. For two years in a row (1931 and 1932), foreign registered visitors exceeded Italian ones – in 1932, almost 60% of the registered visitors were from abroad.

Yet charts 1–4 show that, as the Museum's 'novelty effect' began to fade away, so did the visitor flow. After the *annus mirabilis* of 1932 (582 registered visitors), the number of visits progressively dropped, eventually reaching its lowest point in 1935, when the Guestbook was signed by 321 people. By February, only 14 people had signed the guestbook, marking a downturn of some 70% and 65% with respect of February 1934 and 1933, respectively.[40] These were discouraging numbers for a museum that relied on ticket sales to supplement its meagre income, and after 1935 things would get even worse than that. In the early 1940s, a visitor, Giuseppe Sammicheli, would address an open letter to a newspaper denouncing how "forgotten" the Museum was. "The Museum is having a hard time surviving," Sammicheli would write. "It is not featured in travel guides, and travel agencies make no mention of it when asked by tourists for suggestions on places to visit."[41]

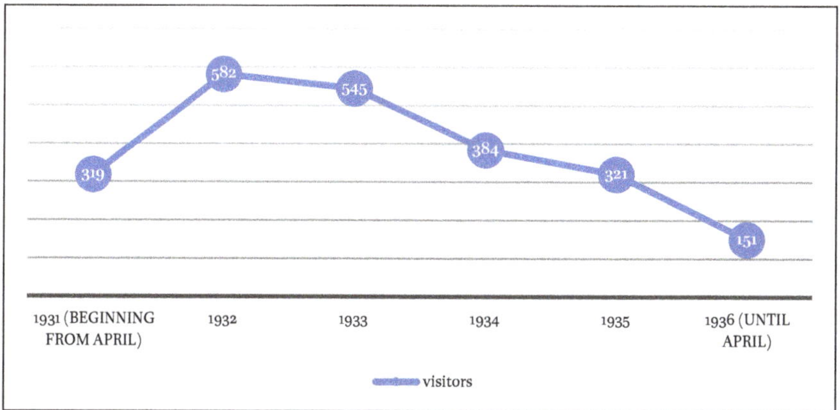

CHART 1 Visitors who signed the guestbook per year

40 *Albo dei visitatori*, vol. I (1931–1953), ArchMG, IMSS.

41 Giuseppe Sammicheli, "Un museo dimenticato," [n.d.], in ArchMG, IMSS, Rit. 39/56.

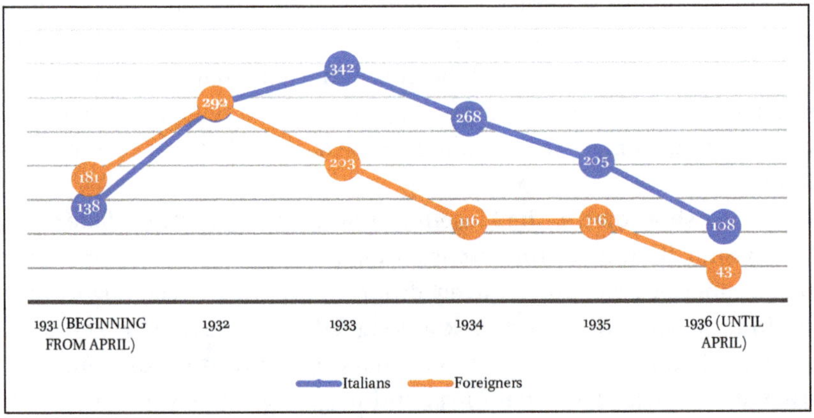

CHART 2 Visitors who signed the guestbook per year

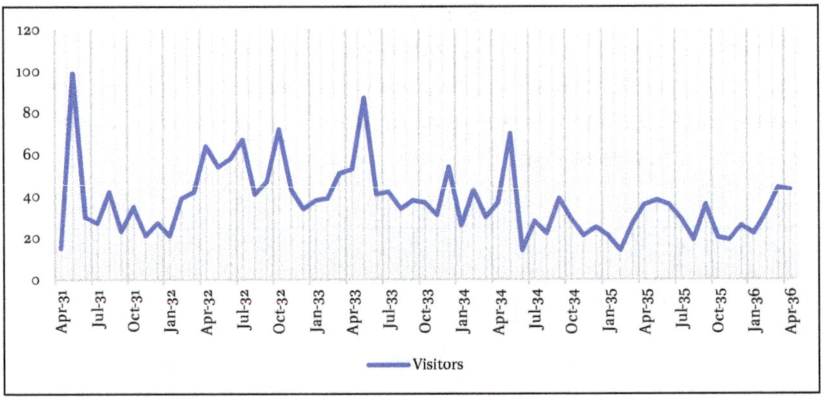

CHART 3 Visitors who signed the guestbook per month

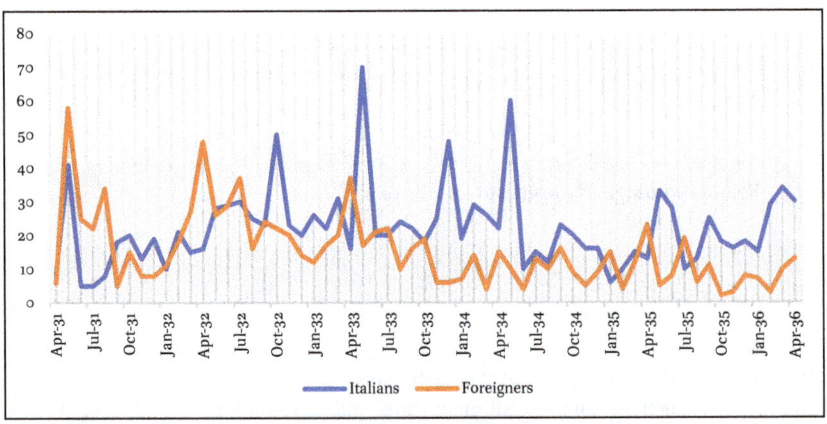

CHART 4 Visitors who signed the guestbook per month

The difficulties encountered by visitors were pretty much the same of those already pointed out by Corrado Sofia during the First Exhibition of History of Science. "Once you get inside," Sofia wrote on that occasion, "you turn to the left and to the right, and you find yourself alone, at the mercy of tags and showcases, looking at imperfectly disposed objects, without a catalogue to guide you."[42] This was very likely true for the Museum, too, as it had no catalogue or guidebook available to visitors, and guided visits were not routinely provided.

Still, Corsini's wish of creating a hybrid institution, an institute/museum that could show to the visitor the historical development of science and could serve at the same time as a research facility for scholars and science lovers, which he had been pursuing since the early 1920s, was taking shape anyway, and its complete realization, according to Corsini, would be the publication of a lavish, comprehensive, and marvelous catalogue of the Italian science.

42 See ch. 4.

CHAPTER 6

Indexing the Italian Scientific Heritage

1 A Definitive Inventory

Announced by the press as one of fascism's "most beautiful initiatives," the so-called *Grande Catalogo-Repertorio di Storia della Scienza* (the Great Inventory-Catalogue of History of Science) was an ambitious project launched in 1928 by the Executive Committee of the First National Exhibition of the History of Science. According to its conceivers, the Great Inventory-Catalogue was to be an index of the Italian scientific collections and, at the same time, a bio-bibliographical directory of Italian scientists. As explicitly stated in the editorial plan drafted in 1928, it was intended as "both an index and a weapon: an index to be peacefully unwinded in front of foreign and local researchers of truth; a weapon to be used against the detractors of Italy and the thieves who ceaselessly try to steal the glorious achievements of the Italian Genius."[1] Thanks to the Great Catalogue-Inventory, the editorial plan went on, "all the indisputable priority claims" made during the Exhibition would receive a definitive confirmation.

This was of course the rhetoric of the Italian Genius we have already come across, only emphasized by the threatening hint at the Great Inventory-Catalogue as a weapon to be used against foreign usurpers; yet behind the thick curtain of propaganda lay a genuine, full-structured cultural program for the development of an effective protection policy. The Great Inventory-Catalogue was, in fact, the revisiting of a project originally launched by Corsini in 1922 and later developed within the National Scientific Heritage Protection Group early in 1924. In the speech delivered at the Bologna meeting of the *Società di storia critica delle scienze mediche e naturali* in 1922, Corsini had indeed remarked that it was necessary to prepare catalogues and inventories; and two years later, following the establishment of the National Scientific Heritage Protection Group, he had urged museums, libraries, and other institutions to join their efforts in order to devise a set of bibliographic and cataloguing standards for indexing historical scientific collections. However, as Corsini's plan to create a "decentered network" of institutions had proved unfeasible due to the cultural and political circumstances, the project had

1 "Relazione sul grande catalogo dell'Esposizione di storia della scienza: programma generale dell'opera," ASCFi, CF9260, fasc. 7.3, 334r.

 | DOI:10.1163/9789004297418_007

been temporarily put aside, and it was only after the Executive Committee of the First National Exhibition of History of Science succeeded in receiving support from the Government that it was revived and put under way.

Since the early 1920s, Corsini had been stressing on the necessity "to undertake in defense of the historical scientific heritage the same action that has been successfully been taken by others in defense of Art." The efforts of people and groups such as Giovanni Battista Cavalcaselle, Giovanni Morelli, and the Tuscan Brigades of the Friends of Monuments had indeed led to the creation, in 1902, of the National Catalogue of the Monuments and Works of Historic, Artistic, and Archaeological Interest, and afterwards, in 1909, to the Rosadi Act itself. Even though they remained far from being fully implemented and suffered from intrinsic limitations, these measures greatly contributed to the identification and preservation of the Italian artistic heritage, laying the foundations for the institutional recognition of the public value of arts and culture. During the 1920s, Corsini worked hard to have the historical scientific collections included in the Government protection policy, stressing on the importance of having a catalogue as complete as possible. In an interview gave to the newspaper *il Giornale d'Italia* on July 27, 1928, he pointed out that

> As for ourselves, we will be happy if the Exhibition succeed in demonstrating the necessity of establishing in Italy the kind of Museums that are being established abroad; if we get a first and large catalogue of the relics and documents related to history of science existing in our nation; if the Government realize that it is time to create, besides the Superintendencies for the Arts, monitoring bodies to watch over the historical scientific heritage.[2]

Corsini's idea is remarkable in that it implies that the Catalogue was to be the fundamental tool for the implementation of an effective protection policy and, at the same time, a large repository of resources for anyone interested in the history of science. Both these traits were summed up in Corsini's insistence on the Catalogue being a "descriptive" work of the kind of Pier Andrea Saccardo's *La botanica in Italia* (published in 1891 and 1901), that is, a repository of materials chronologically arranged by discipline and complete with exhaustive indexes. In a sense, the Catalogue was the image of the Exhibition as it should have been had the local committees not asked for a geographical order. Moreover, it was to portrait the Exhibition as if the same local committees had sent to Florence all the collections they owned. From this perspective, it is

2 "L'Esposizione di storia della scienza," *il Giornale d'Italia*, July 27, 1928.

clear why a plain catalogue of what had actually been on display was not published: for Corsini, the physical exhibition was merely a collection of instances exemplifying the richness of a scientific heritage whose full extent was to be accounted for in the Great Inventory-Catalogue.

But the Catalogue was pivotal for the newly-instituted Museum itself, too, because accomplishing such a project would have meant, for the Museum, crediting itself with being the leading Italian institution devoted to history of science and the main research center for anyone interested in that field of study. In its being both an exhaustive index and a basic tool for research, the Catalogue would have reflected the Museum's twofold commitment to conservation and scholarly research, thus substantiating the Florentines' primacy claims in front of analogous initiatives, were they to take place in Rome or in Milan.[3] Given this, accomplishing the Catalogue became the museum's main concern throughout the 1930s – in fact, apart from the participation in the Chicago World's Fair, it would be the only large-scale project the Museum would focus on. As the Inventory-Catalogue was never carried out, looking back at its bumpy history allows us to better appreciate the ambitions and limits of Italian cultural initiatives during the fascist rule in general and of Corsini's action in particular.

2 The Original Plan

The task of elaborating an editorial plan and carrying out the editorial work was assigned to Umberto Repetti, a young priest descendant of the historian and geographer Emanuele Repetti. At that time, Repetti, a pharmacy graduate of the University of Pisa, was in his late twenties and worked the night shift at the Santa Maria Nuova Pharmacy, reportedly because after having been involved in some unspecified controversy with his ecclesiastical superiors he had been prevented from celebrating Mass. In 1927, he had joined the Institute of History of Science as voluntary assistant on recommendation by Ludovico Granata, his former professor at Pisa and Corsini's friend, and it was Corsini who, the following year, suggested his name to the Executive Committee of the Exhibition as the editor of the Catalogue despite Repetti's lack of editorial experience and his limited publication record.

On March 24, 1928, the Executive Committee made an official proposal to Repetti in a letter signed by Ginori Conti and Guido Del Beccaro. It required the editor to perform all the historical researches deemed necessary for the

3 See below.

purpose, maintain the correspondence with the involved local committees, take care of proofreading and editing the texts, and prepare a short guide of the Exhibition. Repetti would receive for his work a monthly stipend of 600£, up to a total amount of 10.000£. The manuscript should be consigned within one year; the Executive Committee would hold all rights related to the Catalogue, while Repetti's work would be duly credited in the Catalogue's preface; finally, the Executive Committee would watch over the entire work and decide about every aspect of it.[4]

Four days later, Repetti wrote back. Though accepting the general conditions of the contract, he asked to slightly change some terms of it. First of all, he claimed full authorship over the Catalogue and wanted his name being impressed on the frontispiece as, according to him, "no one reads prefaces." Secondly, he was willing to engage in historical researches, provided that they could be carried out in Florence. Finally, he made it clear that he would take care of the text only, leaving to the Executive Committee the encumbrance of finding images, tables, etc. to accompany it and adding that in any case he was not going to proofread the manuscript.[5] The Executive Committee never replied to these requests either in the affirmative or in the negative, giving rise to a number of disagreements between Repetti and the General Secretary of the Exhibition that would soon endanger the entire project. At the beginning of April, however, Repetti set to work despite the pending contractual issues. He began collecting biographical and bibliographical materials, sketched an editorial plan, and drafted a preliminary report about the Catalogue's scope and structure. He presented all these to the Executive Committee during a meeting held on October 17.

On the whole, Repetti's hierarchical plan mirrored the Exhibition's original, disciplinary arrangement. Once gathered, contents were to be organized into five disciplinary sections, i.e. History of Science in General; Natural Sciences; Mathematical, Physical, and Chemical Sciences; Technology. Each one of these sections was divided into several subsections, which in turn contained one of more additional typological categories. For example, the subsection *Chemistry* within the section *Mathematical, Physical, and Chemical Sciences* featured the typological categories *History, Teaching, Treatises etc*; *Instruments, Models, Machines*; and *Portraits, Relics, Biographies*. This structure, however, was not applied uniformly. The section *History of Science in General* had no subsections at all, whilst the subsections *Anthropology, Mineralogy, Paleontology, Pharmacy*, and *Veterinary* contained no typological categories. Moreover, the

4 Ginori Conti and Del Beccaro to Repetti, March 24, 1928, ASCFi, CF9260, fasc. 7.3, 325r–325v.

5 Repetti to Ginori Conti, March 28, 1928, ASCFi, CF9260, fasc. 7.3, 326r.

only subsection of *Technology* was *Engineering*, and the arrangement of the categories within this subsection was not typological, but instead followed a disciplinary logic of its own. A wide apparatus of indexes, devised to help readers find what they were looking for, completed the plan.[6]

The Catalogue, Repetti explained, would feature descriptions of "all the manuscripts, instruments, machines, models, relics, etc. existing in Italy, no matter if they have been on display at the Exhibition."[7] Moreover, Repetti planned to attach to each entry a biography and, where possible, a portrait of the scientist(s) the described object was linked to. As for the indexes, he estimated that they would consists of at least 10.000 entries grouped in four categories: (I) an alphabetically ordered biographical index of scientists; (II) the same, but arranged geographically; (III) an alphabetical index of the items described in the first part of the Catalogue; (IV) a chronological index of the same items. As one may easily see, it was a very ambitious and wide-ranging project – the kind of project, in fact, that would have very likely required several years to be accomplished. Still, incredibly enough, Repetti thought it possible to complete the task in one year, and perhaps more incredibly the Executive Committee considered this to be a realistic deadline. Thus, Repetti's plan was approved without hesitation and the editor set to work under the supervision of the General Secretary, Raffaello Bacci.

A member of the City of Florence Office for Fine Arts and Antiquities since its foundation in 1907, Bacci had been involved in all the major exhibitions held in Florence until then. He had never approved of the Executive Committee's decision of appointing Repetti as managing editor of the Catalogue, and, as Repetti was put under his supervision, the relationships between the two of them soon grew exceedingly cold. In September 1929 Bacci accused Repetti of lack of diligence in carrying out the work; in December the situation escalated until an open conflict broke out. According to the agreement stipulated between Repetti and Bacci, the Secretary would forward data about historical scientific collections as they were received from the local committees to the editor; the editor, then, would perform further research in order to turn the data into actual Catalogue entries. Were these researches proved fruitless, the editor would have filed a request for additional information to be sent by the Secretary to the local committee which had originally sent the data in question. As this routine went on, Bacci noticed that the number of requests being filed by Repetti was exceedingly high, and came to suspect that the editor was not even

6 "Piano generale dell'opera," ASCFi, CF9260, fasc. 7.1, 289.

7 "Relazione sul grande catalogo dell'Esposizione di storia della scienza: programma generale," ASCFi, CF9260, fasc. 7.3, 331r.333v.

attempting at researching as he was supposed to do by contract, but simply filed requests for additional information as soon as he received the data from the Secretary. The situation reached its boiling point in December, when Bacci received from the University of Cagliari a note signaling "34 wood boxes containing anatomical waxes made in Florence by Susino and his students."[8] The General Secretary transmitted the note to Repetti, who filed a request for information about "Susino." This request, once signed by Corsini, was forwarded to the University of Cagliari, whose reply turned out to be mortifying. It was funny, the Sardinian University pointed out not without sarcasm, that the Florentines had asked about Clemente Susini, himself a Florentine, and quite a renowned one. Upon reading the letter, Corsini remonstrated with Bacci, who then turned to Repetti:

> Dear Mr. Repetti,
>
> I am sorry to remind you that, according to our agreement, it is you who are supposed to perform all the biographical researches for the Catalogue, not the local committees.
>
> We can rely on the local committees' courtesy only if we happen to deal with lesser-known scientists, about whom carrying out researches here in Florence would be too long and laborious [...]. Now, asking the local committees for information about famous scientists who lived and worked exclusively in Florence, as you did for CLEMENTE SUSINI, is clearly not the case. [...]
>
> I do not know what biographies the local committees are going to provide, but mind: no one must be missing from the Catalogue.[9]

Repetti replied to these complaints by stating that the notes sent by local committees were often incomplete or inaccurate: in the specific case of Susino/Susini, he explained that he had no such biographical knowledge as "to divine that one 'SUSINO' ... was actually one 'CLEMENTE SUSINI' who lived and worked exclusively in Florence."[10] For Bacci, however, this was the proof that Repetti was not putting enough care in the job. "If you had looked at the objects listed in the note," the Secretary complained, "you would have realized, even without possessing the biographical knowledge you say you wish you owned, that SUSINO was just a little typo for SUSINI." And this was not even the first case, Bacci insisted, for there had been many others. "Be careful, Mr. Repetti,"

8 ArchMg, Esp29, "Schedario delle segnalazioni – Cagliari."

9 Bacci to Repetti, December 19, 1928, ASCFi, CF9260, fasc. 7.3, 327r.

10 Repetti to Bacci, December 19, 1928, ASCFi, CF9260, fasc. 7.3, 327r.

he concluded, "and pay attention to these little things, as we live in a world full of people with planted guns ready to hurl at us bittersweet remarks like the ones prof. Corsini received from Cagliari and Cremona."[11]

Bacci did not fail to inform Ginori Conti about the situation and particularly about Repetti's poor behavior. The Prince, worried that this unfortunate turn of events would delay the completion of the Catalogue, summoned Repetti to appear and report about his work before a "special Commission for the Catalogue" consisting of Ginori Conti, Abetti, Gino Bargagli Petrucci, Corsini, Angelo Bruschi, and Giuseppe Boffito. During the meeting, held on January 15, 1929, Repetti presented to the Commission the work he had done. From the 450 descriptive notes he had by then received from the local committees, he had drawn 2400 records for the analytical index and 1330 scientists' names. He had sketched out 330 out of 830 biographies and asked the local committees for the remaining 500. Once the local committees had sent all the information about their collections and the analytical index was finished, he would set to work on the other indexes and on the missing biographies. Therefore, he concluded, the deadline for the consignment of the catalogue should be extended until August and the editor be assigned a proper office and an assistant.[12]

On January 19, the Commission communicated its resolutions to Repetti. Upon close examination, the Commissioners had found Repetti's work barely acceptable: the editor, they observed, should have maintained all the indexes up to date as the descriptive notes were coming in and the same should have been done for the Catalogue entries; as for the biographies, they did not fulfil the Catalogue wished standards, as they were, according to the Commissioners, "concise to the point of giving not even the remotest idea of the life and works of the scientists they are about." The Commissioners had hoped that at least some section of the Catalogue was complete, but after having seen what Repetti had done they had to conclude that "nothing, up to this point, is in the condition of being sent to the printmaker."[13]

The Commission, however, accepted Repetti's proposal to extend the deadline until August, but, in order to avoid the kind of problems denounced by Bacci, decided that Repetti would work at the National Library under director Angelo Bruschi's surveillance.[14] So, on January 26, Repetti resettled at

11 Bacci to Repetti, December 20, 1928, ASCFi, CF9260, fasc. 7.3, 329r.

12 "Relazione sul grande catalogo dell'Esposizione di storia della scienza: programma generale dell'opera," ASCFi, CF9260, fasc. 7.3, 332r.

13 Ginori Conti to Repetti, January 19, 1929, ASCFi, fasc. 7.3, 335r.

14 Ginori Conti to Repetti, January 19, 1929, ASCFi, fasc. 7.3, 337r.

the National Library, where he was provided with an office and was granted unrestricted access to the Library holdings. On the next day Bacci forwarded to Bruschi all the documentation related to the Catalogue. Repetti, however, was just a few days into his work when he was informed that Giuseppe Boffito was to take over the entire project and he would serve as editorial assistant to the new editor.[15]

3 The Revised Plan

During the meeting of January 15, Boffito had proposed to replace the "biographical index" of the Catalogue with a "bibliography on scientific instruments" which methodically listed all the books and papers dealing with individual scientific instruments, so that, for example, readers could easily find bibliographic sources on telescopes, solar clocks, sextants, astrolabes, and so on. "Such a bibliography," Boffito argued, "would be very helpful and would fill a gap in our discipline."[16] Corsini judged Boffito's proposal to be "very palatable and appropriate" but thought it unnecessary to change the Catalogue's general plan, for Boffito's bibliography could suitably be added to the work as a brand-new section, provided that Boffito himself was willing to take care of it and, at this point, of the entire Catalogue as well.[17] So, an agreement having been duly signed, on the 3th of February Boffito officially took on the directorship of the whole project.[18]

Boffito – a sixty years old Barnabite Father of the *Collegio alla Querce* of Florence – was a skilled polymath, having written extensively on Dante Alighieri, Cecco d'Ascoli, the history of Aviation, and the history of the Barnabite Order. In 1929, he had just completed a pivotal work on the history of scientific instruments, *Gli strumenti della scienza e la scienza degli strumenti* ("the Instruments of Science and the Science of Instruments"), a book dedicated to Piero Ginori Conti and published in conjunction with the Exhibition.[19]

The presence of Boffito at the head of the project made the Commissioners feel assured and confident that the Catalogue would be published in time. In March, the Executive Committee of the Exhibition began negotiating a deal with the publishing house *Rinascimento del Libro* ("Book Renaissance"); and

15 Ginori Conti to Repetti, January 31, 1929, ASCFi, fasc. 7.3, 338r.
16 Boffito to Corsini, [n.d.] ASCFi, CF9260, fasc. 7.3, 347r.
17 Corsini to Boffito, January 23, 1929, ASCFi, CF9260, fasc. 7.3, 347r.
18 Boffito to Ginori Conti, February 3, 1929, ASCFi, CF9260, fasc. 7.3, 350r.
19 On Boffito, see the entry in *DBI* by Ada Alessandrini.

on the opening day of the Exhibition Ginori Conti officially announced the forthcoming publication:

> A GREAT CATALOGUE OF HISTORY OF SCIENCE will be published soon [...]. It will show the importance of the Italians' contributions to the development of modern civilization throughout the centuries. This Catalogue – properly illustrated and enriched by over three thousand biographies, and analytical, topographical, and disciplinary indexes – is bound to stand as the most important reference book for historians of science.[20]

By that time, however, little had been done. Besides the fact that Repetti, keeping faith to his resolution to take care of the Catalogue's text only, had done nothing about the pictures, the editorial work was going on very slowly – in fact, on the opening day of the Exhibition, the entire project was already in stand-by status due to Repetti's commitment to working on the Exhibition Guide Book, a task that would keep him busy until the end of June. Moreover, right after taking office, Boffito had Repetti work on the new bibliographic section and leave aside the work on the indexes and biographies; he also asked Repetti to rewrite all the Catalogue entries and classify them according to the provenance of the items described therein, dismissing the thematic classification Repetti had been adopting up to that point. Repetti thus spent the remaining months of his contract working on the Guide Book and managing the information coming from the local committees following Boffito's instructions. When the August deadline came, he had classified all the Catalogue entries, completed the analytical index, and sketched out 1300 biographies; yet the Catalogue itself was still far from being ready to be printed, all the more so because in the meanwhile Boffito had completely changed Repetti's original plan.[21]

In his capacity as managing editor of the Catalogue, the Barnabite Father thought he could freely modify the original plan without any prior consultation with the Commissioners. Thus, he had crossed out both the five disciplinary divisions and the indexes, and replaced them with his own plan, according to which the Catalogue was to be divided into three major sections. The first one would consist of the Catalogue entries arranged geographically in order to show "the Italian cities and regions in their historical relationships with science"; the second, of the "Bibliography on scientific instruments"

20 "Il re ha inaugurato a Firenze la Mostra nazionale di storia della scienza," *il Giornale d'Italia*, May 9, 1929.

21 Repetti to Corsini, October 1, 1929, ASCFi, CF9260, 339r.–343r.

devised by Boffito; and the third, of the alphabetically ordered biographical Index prepared by Repetti. This plan, Boffito assumed, emphasized bibliography, and therefore complied better with the Commission's wish of publishing a reference work.

Corsini found out about Boffito's new plan only in September, when, in the imminence of the opening of the SIPS annual meeting, Boffito had *Rinascimento del Libro* print brochures describing the Catalogue's scope and structure to be handed out among the meeting participants. Before printing the brochures, *Rinascimento del Libro* sent out a proof for approval to Corsini, who, after reading it, stopped the publisher from printing them and informed Ginori Conti of what the whole thing was about. "Boffito's plan (which I am attaching to this letter)," he wrote, "does not correspond to the one we agreed on."[22] What discontented Corsini the most was that Boffito's plan reflected the Exhibition as it actually was and not, as Corsini wished, as it would have been had the local Committees not stood in the way of it. Furthermore, he believed that Boffito's emphasis on bibliography clashed with the Catalogue's ambition of being a purely "descriptive" work – even though it is hard to tell what Corsini actually had in mind when he reported to the Prince about Boffito drawing up an "interpretive" work. Worried about this unexpected turn of events, Ginori Conti called a meeting of the Catalogue Commission to be held on the 6th of October.

Meanwhile, the relationship between Repetti and Bacci had not improved, notwithstanding the fact that the former had been working under Boffito's and Bruschi's supervision. In a letter sent to Corsini on the 1th of October, Repetti complained about the lack of support from the Secretary, who would send the local committee's notes over to him at an exasperatingly slow pace.[23] In another letter sent on October 4 to Ginori Conti, he explained that he had followed Boffito's instructions accurately, and therefore could not be reasonably held responsible for the delay in the completion of the Catalogue.[24] Bacci, for his part, after having been informed about Repetti's complaints, on the very day before the meeting addressed a long and detailed report to Ginori Conti recapitulating the events and blaming Repetti for having failed to deliver the Catalogue in time. "Thus Dr. Repetti," Bacci maintained at the end of his missive, "MUST BE COMPELLED TO COMPLETE THE CATALOGUE without any further payment."[25]

---

22 Corsini to Ginori Conti, September 30, 1929, ASCFi, CF9260, fasc. 7.1, cc.299r–300r.

23 Repetti to Corsini, October 1, 1929, ASCFi, CF9260, 342r.

24 Repetti to Ginori Conti, October 4, 1929, ASCFi, CF9260, fasc. 7.3, 344r–345r.

25 Bacci to Ginori Conti, October 5, 1929, ASCFi, CF9260, fasc. 7.3, 316r. Emphasis is in the original.

At the meeting of October 6, the Catalogue Commission gathered at Ginori Conti's house and decided that the work for the Catalogue should go on according to the original plan, whilst Boffito would continue to work on the new bibliographic section about scientific instruments. Repetti would reacquire his position as managing editor and be granted another year of stipend support; finally, both Boffito and Repetti would benefit from the aid of Pietro Pagnini and Riccardo Grassina. With these provisions the Commission hoped to have the work finished by the end of 1930, all the more so because during the meeting it was also announced that the Florence Savings Bank had decided to sponsor the project by donating £40.000.

The year 1930 passed with Repetti collecting and editing biographies and catalogue entries, Boffito gathering bibliographical resources on scientific instruments, and Pagnini and Grassina checking the notes received from the local committees and drawing up an inventory of the items on display at the Exhibition. Corsini, Ginori Conti, and Bacci, absorbed as they were in preparing the final accounts of the Exhibition and establishing the Museum, would rarely inspect the work done for the Catalogue. Come December, Repetti delivered the manuscript of the Catalogue to the Commission: it consisted of 1260 typewritten pages ready to be forwarded to the printer. As for Boffito's bibliography, on November 30 the Barnabite Father had written to Ginori Conti, explaining that the work was still far from being complete.[26] Since neither the Catalogue Commission nor the Museum could afford to cover the costs of the bibliography any more, it was decided to suspend Boffito's work and go ahead with the printing of the Catalogue. So, at the end of 1930, Bargagli Petrucci began negotiating a deal with *Rinascimento del Libro* for the publication of the work.

4 Editorial Troubles

Rinascimento del Libro was a small publishing house founded in 1926 that was trying to make its way into the publishing market by means of fine editions of classics and art books. Its founder and executive manager, Enrico Barfucci, had been one of the promoters of the creation of the EAT and a member of the Executive Committee of the First National Exhibition of History of Science. The first book published by *Rinascimento del Libro* had been Giacomo Leopardi's *Canti* (1927), which inaugurated the "National Collection of Italian Classics" series edited by Florentine writer Giovanni Papini. By 1930, *Rinascimento del Libro* had also published two large-sized, richly illustrated art catalogues,

26 Boffito to Ginori Conti, November 30, 1930, ASCFi, CF9260, fasc. 7.2, 295r.

the *Atlas of Modern Engraving* and *The Romanic Sculpture in Tuscany*. The Catalogue of History of Science was intended as the first in a brand-new "History of Science" series edited by Garbasso (the series was never realized, for Garbasso died in 1933).

Rinascimento del Libro, in concert with Corsini, appointed Antonio Gigli – the archivist and librarian of the count Guicciardini – to check and proofread Repetti's texts. Gigli received them in May 1932, and upon scrutinizing the manuscript realized how poorly edited it was. The badly-written texts needed to be reworked and abbreviations spelled out; typos abounded and dates would often turn out to be inaccurate; illustrations were missing and indexes rather mixed up. Barfucci, being asked to indicate a possible release date, explained to Ginori Conti that the manuscript needed so much work that it was nearly impossible to have the Catalogue published before the middle of 1933.[27] Yet in June of that year, as the deadline approached, he requested a meeting with the Prince to discuss the matter thoroughly, for he was convinced that, given the conditions of Repetti's manuscript, they ought to think about changing the Catalogue's structure altogether. "Gigli is absolutely right," Barfucci stated. "Repetti's manuscript is truly awful."[28] According to him, the most feasible thing to do was to catalogue what had been actually on display, letting the other things go. If the Prince did consent to that, Barfucci argued, the Catalogue would have been published "relatively soon." Even though it is impossible to tell whether or not the meeting between the two actually took place, it is apparent from Corsini's preface to the Catalogue drafted in March, 1934, that Barfucci's proposal was accepted:

> At the time of the Exhibition, a brief Guide, roughly describing the items on display grouped by their provenance, was published to assist the visitors during their visits. Back then, we promised to publish shortly thereafter a Catalogue-inventory of history of science featuring the items which were on display and all the others that the local committees had informed us about, and complete with biographical and bibliographical sketches of leading Italian scientists. Such publication was delayed for several reasons, but mainly because we decided to direct our efforts to provide a proper seat and accommodation to the National Museum of History of Science of the University of Florence [...] and because the Treccani Encyclopedia had meanwhile decided to compile a Dictionary

27 Barfucci to Ginori Conti, November 2, 1932, ASF, PGC archive, s. 1, fasc. "Tipografie e stamperie."

28 Barfucci to Ginori Conti, June 23, 1933, ASF, PGC archive, s. 1, fasc. "Tipografie e stamperie."

> of Italian Biography. [...] So we thought it better to publish the complete catalogue of what had been on display at the Exhibition, postponing the publication of a more general catalogue until the Museum becomes a National Museum in the truest sense of the word, so that will be possible to gather data, information, and instruments from all over Italy.[29]

Though Corsini omitted to mention editorial troubles among the reasons that had actually delayed the publication of the Catalogue, it was true that there were initiatives that, being better funded and managed, would make some aspects of the Catalogue look mediocre – if not amateurish – in comparison. The Dictionary of Italian Biography, for example, was a project conceived by Gentile in 1925, and though its first volume would be published only in 1960, in 1934 it was receiving strong support from the Government by means of the *Istituto dell'Enciclopedia Italiana*. Its managing editor, the bibliographer and Gentile's collaborator Fortunato Pintor, planned to gather documents and information about the life and works of any Italian mentioned in or related to official State documents and works of literature, science, and art, involving in this as many experts as possible.[30] In 1932, the newspaper *il Popolo d'Italia* announced that an awe-inspiring work in twelve volumes, titled "The works of the Italian Genius around the world" and consisting of biographies of Italian scientists, artists, travellers, warlords, politicians, architects, Saints, and merchants, was being prepared in Rome.[31] Around the same time, Giulio Provenzal, backed up by the CNR and expanding on the work done for the Chicago's World Fair of 1933, was working on a "Documentary collection of the Italian scientific primacy," which in his intention would serve as the basis for the establishment of a National Museum in Rome.[32] Furthermore, while all these were happening, engineer Guido Ucelli had succeeded in creating a "Commission for the establishment of a Science and Industry Museum" within the Milan City Council, the Museum itself being a project supported by Belluzzo, approved by Marconi, and longed for by Mussolini.[33] In Florence,

29 "Prefazione del Grande Catalogo-Repertorio," draft, 1934, ASF, PGC archive, s. 2, fasc. "storia della scienza (1)," 1.

30 See Roberto Pertici's introduction to *Giovanni Gentile e il Senato. Carteggio (1895–1944)*, edited by Emilia Campochiaro, Lucia Pasquini, and Alessandra Millozzi (Soveria Mannelli: Rubbettino, 2008).

31 "Italiani 'esportatori di civiltà'" *il Popolo d'Italia*, March 2, 1932.

32 See Stefano Canali, *La medicina*, in *Per una storia del Consiglio Nazionale delle Ricerche*, vol. I., edited by Giovanni Paoloni and Raffaella Simili (Roma-Bari: Laterza, 2001).

33 See Guido Ucelli, *Origine realizzazioni e programmi del Museo Nazionale della Scienza e della tecnica "Leonardo da Vinci"* (Milano: s.n., 1955).

initiatives as such were considered insidious at best: "Dawdling over the publication of the Catalogue and so let analogous initiatives overcome us would be very disgraceful," Ginori Conti wrote to Bacci in 1934.[34]

In any case, the decision of publishing a catalogue featuring only what had been on display, leaving aside biographies and indexes, did not turn out as rewarding as the Prince and Corsini had expected. Since the end of 1933, Gigli had been submitting batches of revised texts to *Rinascimento del Libro*'s subsidiary company *Tipocalcografia Classica*, the publishing and printing facility managed by Bacci which was now entrusted with the task of printing and publishing the Catalogue. And in November, 1934, Bacci wrote to Ginori Conti to update him about the status of the project. The Catalogue was taking so much time to be completed, Bacci wrote, because of Repetti's incompetence ("that blessed unfrocked priest") and Gigli's excessive diligence ("I am clashing with a very hard rock, that is, prof. Gigli's carefulness [...], for he is a very meticulous and precise man."), but also, and above all, because of Corsini's lack of foresight in choosing them for the job. "It was not I who chose Gigli," Bacci insinuated. "As it was not I who chose Repetti." However, Bacci concluded, 120 pages had already been printed, and if Ginori Conti was willing to put pressure on Gigli, the Catalogue would be published very soon.[35]

Still, when in spring 1935 Ginori Conti turned to Gigli and solicited him to speed up the work, the latter responded that he was carrying on a task "he had neither sought nor desired," and for which he had received no payment so far. If his work was not appreciated, Gigli went on, he was ready to resign from his job without claiming any money.[36] So Ginori Conti called on Bacci to pay Gigli, but Bacci insisted on Gigli being paid only after the delivery of the complete manuscript. "The texts have been coming in in dribs and drabs," Bacci stated; hence he was afraid that if Gigli got his money beforehand, "they would incur the same misfortunes they had already experienced while dealing with Most Reverend Father Repetti."[37] Both Gigli and Bacci, however, assured the Prince that the Catalogue would be published by the end of the year.

In January, 1936, the Prince, having heard nothing about the Catalogue, asked Bacci for updates. The Secretary blamed Gigli for the delay and suggested to remove him from office.[38] Gigli, as he knew about Bacci's charges, on

---

34 Ginori Conti to Bacci, [n.d.], 1934, ASF, PGC archive, s. 2, fasc. "storia della scienza (2)."

35 Bacci to Ginori Conti, November 13, 1934, ASF, PGC archive, s. 2, fasc. "storia della scienza (1)."

36 Gigli to Ginori Conti, July 4, 1935, ASF, PGC archive, s. 2, fasc. "storia della scienza (1)."

37 Bacci to Ginori Conti, November 13, 1934, ASF, PGC archive, s. 2, fasc. "storia della scienza (1)."

38 Bacci to Ginori Conti, January 13, 1936, ASF, PGC archive, s. 2, fasc. "storia della scienza (1)."

January 20 resigned on his own initiative and without advancing any request of payment for the job he had already done.[39] Then Ginori Conti, rather puzzled and annoyed at Bacci's demeanor, requested the Secretary to name someone who suited him in order to close the matter once and for all.[40] Bacci chose Pietro Pagnini, the man whom he would have appointed already in 1928, had Corsini not decided to give the job to Repetti instead. Pagnini, an expert on scientific instruments, had led tours of the Exhibition, compiled the inventory of the items on display, and in 1936 was informally employed as curator of collections at the Museum of History of Science.[41]

In the following months, Pagnini regularly submitted parts of the revised manuscript to *Tipocalcografia Classica*, whilst Bacci would reassure Ginori Conti that the printing process was going on "at the double," and that the publication was "very imminent." Yet towards the end of August, 1936, after Pagnini announced that the last batch of text had been sent to *Tipocalcografia Classica*, any attempt to reach Bacci ended up fruitless. Corsini and Ginori Conti wrote several letters to the Secretary receiving no reply, and by the end of 1936 they knew nothing about the fate of the publication. Hence, at the beginning of January, 1937, they decided to go to *Tipocalcografia Classica*'s office in person: they met there with the new executive manager, Guglielmo Vita, who regretfully informed them that Bacci had withdrew from the office on October 5, 1936, taking with him all the materials related to the Catalogue, both manuscript and printed, and that therefore the commitment taken on by *Tipocalcografia Classica* could not be honoured.

5 The End of the Project and the Catalogue of 1952

We do not know why Bacci withdrew from office, and we do not know why he decided to take the Catalogue away with him or why he did not reply to Corsini's and Ginori Conti's letters either. Certainly, something might have happened – some kind of disagreement or argument, maybe, for on January 23, 1937, Ginori Conti's lawyer, Guido Del Beccaro, informed the Prince that his attempts to find an "amicable agreement" with Bacci had failed. "Bacci is

39 Ginori Conti to Corsini, May 21, 1936, ASF, PGC archive, s. 2, fasc. "storia della scienza (1)." Attached to Corsini's letter there is Gigli's resignation one, dated January 20.

40 Ginori Conti to Bacci, January 21, 1936, ASF, PGC archive, s. 2, fasc. "storia della scienza (1)."

41 Pagnini, a declared anti-fascist, was prevented from working at public institutions. At the time, he would teach at religious schools.

clearly stonewalling," Del Beccaro reported. "And I would never have believed he could act this way. [...] Hopefully I shall soon be able to report to you in more detail, so that we can decide on the most appropriate court action to be taken against him."[42]

Taking a court action against Bacci, however, was not without risks and the outcome of it could not be taken for granted. Bacci was indeed entitled to author's rights over the Catalogue according to the agreement reached between *Rinascimento del Libro* and the Catalogue Commission in 1932, and thus it was debatable whether he could be forced to hand the manuscript over to the publisher – who, according to the same agreement, was granted commercial rights only – or even to the Executive Committee of the Exhibition itself, as the latter officially appeared just as promoter and supporter of the publication. Moreover, the situation was further complicated by the fact that Bacci was at the same time General Secretary of the Exhibition, author and managing editor of the Catalogue, and a member of *Tipocalcografia Classica*'s board of administrators.

It is unclear whether a lawsuit against Bacci was actually filed. On February 2, Del Beccaro was summoned at Ginori Conti's house to discuss the matter, but there is no extant documentation about the meeting. It is certain that Bacci never consigned either the manuscript of the Catalogue or the administrative records of the Exhibition he had been entrusted with while serving as General Secretary, but no more is known about him besides the fact that he was then serving as General Secretary of the *Mostra Giottesca*.[43] The matter of the Catalogue resurfaces in the Museum records for the year 1938 as the first point being debated at the Museum Board meeting of January 5, 1938. The official minutes, drafted by Corsini, read:

> Lastly, the president communicates to the aforesaid mr. Vita the following deliberation: the publisher, hiring whomever he considers fit for the purpose, must present to the Museum Board a definitive plan for the Catalogue; afterwards, namely, once the plan is approved by the Museum Board, he must present all the drafts, and commit himself not to publish the Catalogue without the approval of the Board.[44]

---

42 Del Beccaro to Ginori Conti, January 26, 1937, ASF, PGC archive, s. 2, fasc. "storia della scienza (1)."

43 On the *Mostra giottesca*, which celebrated the 600th anniversary of the death of Giotto, see Alessio Monciatti, *Alle origini dell'arte nostra. La* Mostra Giottesca *del 1937 a Firenze* (Milano: il Saggiatore, 2010).

44 Minutes of the meeting held on January 5, 1938, ArchMG, "Verbali IMSS, 1938–1939."

So, it does appear that, at the beginning of 1938, the Executive Committee decided to restart the whole project all over again, renouncing even to the diminished catalogue which had been approved jointly with Barfucci and *Rinascimento del Libro* in 1934: it was now up to *Tipocalcografia Classica* to present a suitable project. Still, all the written solicitations sent to Vita in the following months went unheard, apparently, and in November the Museum Board was still waiting for *Tipocalcografia Classica*'s project, as it does appear from the minutes of the meeting held on November 21, 1938.[45] Then in January, 1939, the Prince wrote to Corsini and asked him to take upon himself the long-standing task. "I haven't heard of the Catalogue anymore," Ginori Conti wrote. "And I am very sorrow about that. Please, see if you can manage to get the thing done because, having already secured mr. Vita's support, we could easily accomplish it."[46] As a matter of fact, this is the last piece of information we have on the Catalogue, for in the following months Ginori Conti's mind would be occupied with graver matters. On February 20, 1939, the fascist Government nationalized the Prince's company by turning it into a joint-stock company whereof almost 90% was owned by the Government itself through its holding company *Ferrovie dello Stato* (State Railway), *de facto* dispossessing the Ginori Conti family.

The Prince passed away on December 3, 1939, at his house in Florence. He was 74-year-old. With his death, the Museum lost his main patron and supporter. His son, Giovanni, who succeeded him as president of the Museum's Board of Trustee, was unwilling to invest any more money in an admittedly undefined, grueling, and money-draining project like the Catalogue, and so everything was idled until further notice. It was only after World War II that the project was resumed by Maria Luisa Bonelli, who had been appointed as Corsini's assistant in 1942. Together with Pagnini, who had officially become the Museum's curator in 1940, Bonelli began working on a catalogue of the instruments and machines that had been on display in 1929. They revised Pagnini's inventory, collected Exhibition cards, and contacted lenders to ask for information about the items' current conditions.

The Catalogue was published in 1952 and marked the beginning of a new phase in the Museum's life. Freed from the weight of the somewhat erratic fascist propaganda and its obsession on centralism, the Museum began to focus on well-defined, better-balanced projects. When Corsini died in 1961, he left to his successor Bonelli a Museum on its way to becoming one of the world's

45 Minutes of the meeting held on November 21, 1938, ArchMG, "Verbali IMSS, 1938–1939."

46 Ginori Conti to Corsini, January 20, 1939, ASF, PGC archive, s. 2, fasc. "storia della scienza (2)."

leading institutions devoted to history of science. In the foreword he wrote for the Catalogue of 1952, he managed to link this new chapter in the Museum's life to the views he had expressed back in 1922 at the Bologna meeting of the *Società di storia critica delle scienze mediche e naturali*. "When, in 1929, benefiting from Prince Piero Ginori Conti's support and patronage, I organized the First National Exhibition of History of Science, my aim was to promote the Italian historical scientific heritage," he wrote. "And this, not only so that such an important wealth could be known and appreciated by the public and the community of scholars, but also so that such a call could provide a spur to save from oblivion and dispersion all the materials which, though now considered outdated and without value, can still allow us to follow the steps of scientific development."[47]

47 Andrea Corsini, *Foreword*, in Maria Luisa Bonelli and Pietro Pagnini, *I. Esposizione nazionale di storia della scienza, Firenze, maggio-ottobre 1929: catalogo con aggiornamenti* (Firenze: Leo S. Olschki, 1952), pp. 3–4: 3.

Plates

FIGURE 1 Andrea Corsini at work in his laboratory during the 1910s
© MUSEO GALILEO ARCHIVE

 | DOI:10.1163/9789004297418_008

FIGURE 2 The *Tribuna di Galileo* at the end of the XIXth century
© MUSEO GALILEO

FIGURE 3 Group photo of participants in the Bologna meeting of the *Società di Storia Critica delle Scienze mediche e naturali*
© MUSEO GALILEO

FIGURE 4 Piero Ginori Conti in Arcetri

FIGURE 5 University of Florence alumni at the Institute of History of Science, 1927

FIGURE 6 Antonio Garbasso and Piero Ginori Conti in Larderello, 1928
© ASDFAA

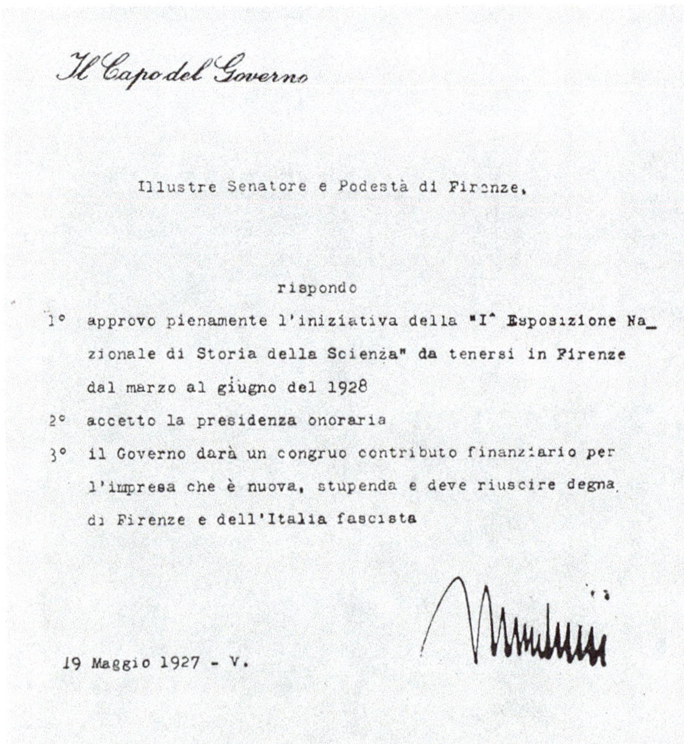

Il Capo del Governo

Illustre Senatore e Podestà di Firenze,

rispondo

1° approvo pienamente l'iniziativa della "I^ Esposizione Nazionale di Storia della Scienza" da tenersi in Firenze dal marzo al giugno del 1928

2° accetto la presidenza onoraria

3° il Governo darà un congruo contributo finanziario per l'impresa che è nuova, stupenda e deve riuscire degna di Firenze e dell'Italia fascista

19 Maggio 1927 - V.

Mussolini

FIGURE 7 Mussolini's message approving the Exhibition
© MUSEO GALILEO

FIGURE 8 Flyer announcing the Exhibition
© MUSEO GALILEO

FIGURE 9 The palace of expositions

FIGURE 10 Vittorio Emanuele III at the Exhibition opening

FIGURE 11 Poster of the Exhibition
© MUSEO GALILEO

FIGURE 12 Pisa exhibit
© MUSEO GALILEO

FIGURE 13 Florence room
© MUSEO GALILEO

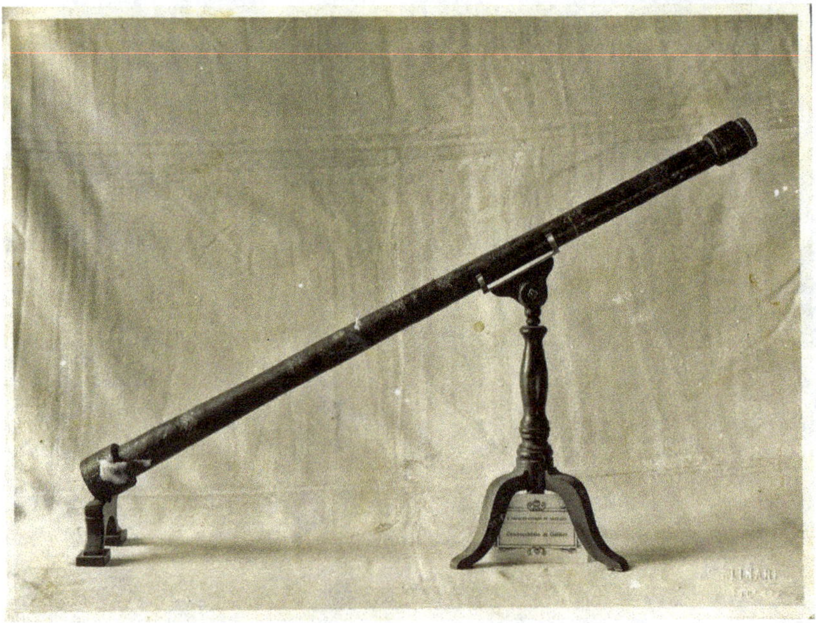

FIGURE 14 Galileo's telescope
© MUSEO GALILEO

FIGURE 15 Bologna room
© MUSEO GALILEO

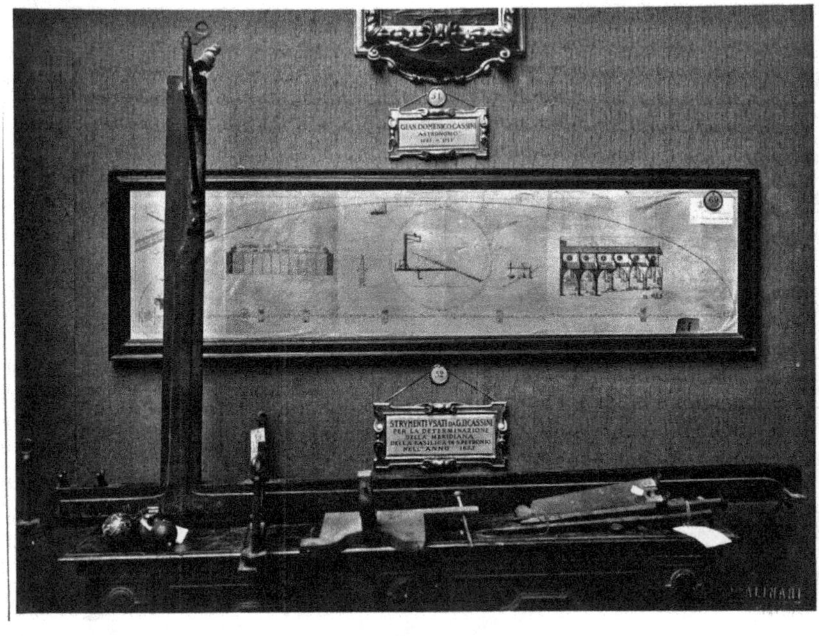

FIGURE 16 Gian Domenico Cassini's instruments
© MUSEO GALILEO

FIGURE 17 Rome room
© MUSEO GALILEO

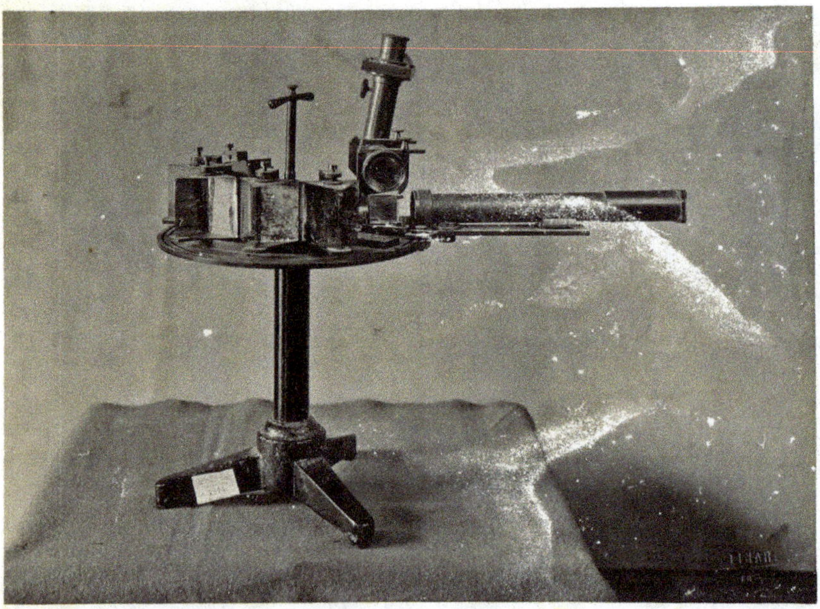

FIGURE 18 Angelo Secchi's spectroscope
© MUSEO GALILEO

FIGURE 19 Milan room

FIGURE 20 Model locomotive

FIGURE 21 Turin room
© ALINARI ARCHIVES

FIGURE 22 Galileo Ferraris's instruments
© MUSEO GALILEO

FIGURE 23
Genoa room

FIGURE 24
The exhibit of Pavia

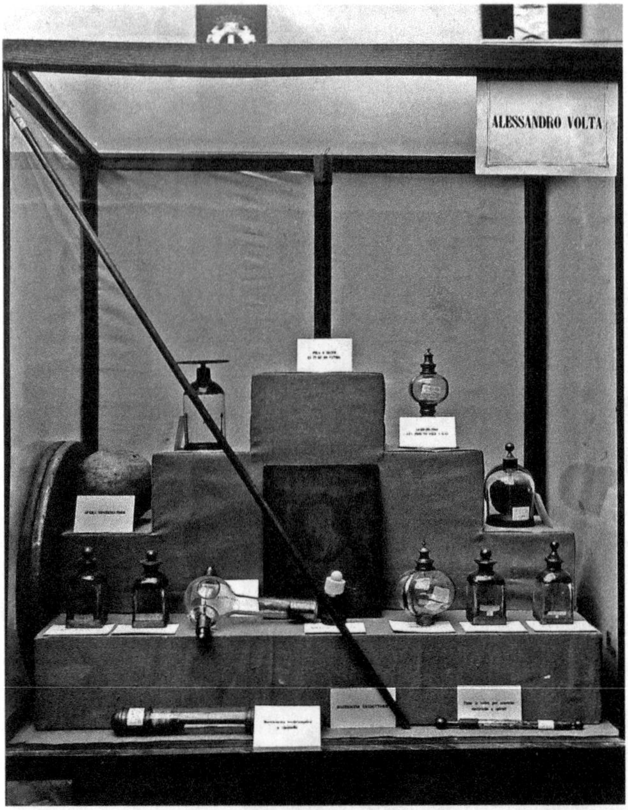

FIGURE 25
Alessandro Volta's instruments

FIGURE 26
The aerology room

FIGURE 27 The aviation pavilion

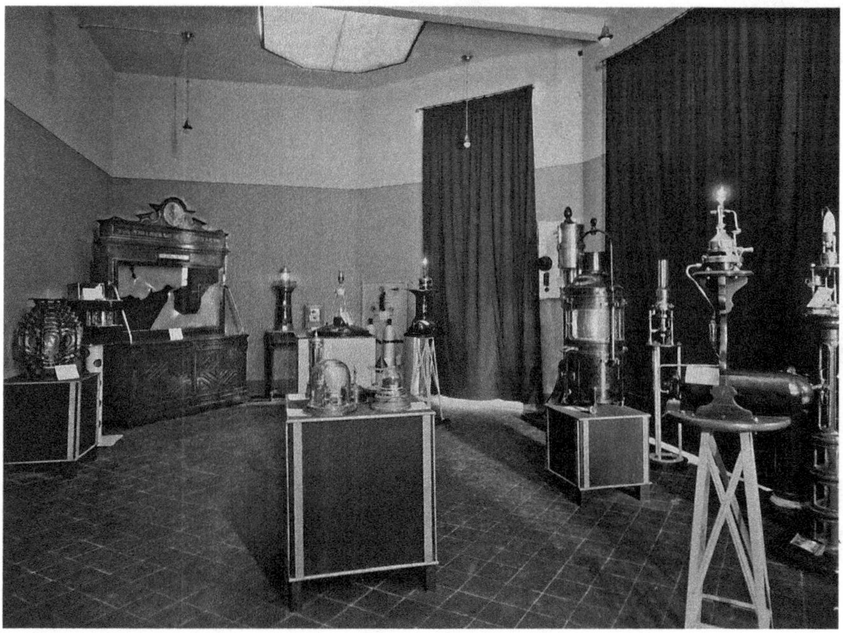

FIGURE 28 The Lighthouse National Service exhibit

FIGURE 29
Model of a wing, after Leonardo da Vinci
© MUSEO GALILEO

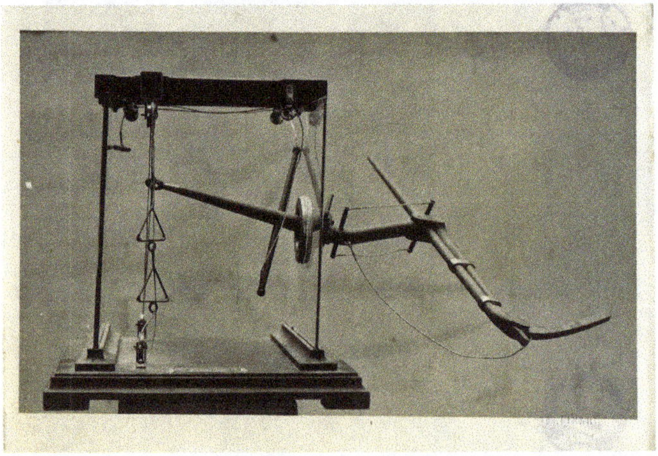

FIGURE 30
Wing models, after Leonardo da Vinci
© MUSEO GALILEO

FIGURE 31 Parachute and propeller, after Leonardo da Vinci
© MUSEO GALILEO

FIGURE 32 Screw lathe, after Leonardo da Vinci

FIGURE 33 Quintino Sella's electromagnetic sorting machine

FIGURE 34 Anatomical preparations by Bartolommeo Pannizza

FIGURE 35 Giovanni Battista Amici's telescope

FIGURE 36 Giovanni Caselli's *Pantelegrafo*
© ALINARI ARCHIVES

FIGURE 37 Palazzo Castellani in the early 1930s
© MUSEO GALILEO

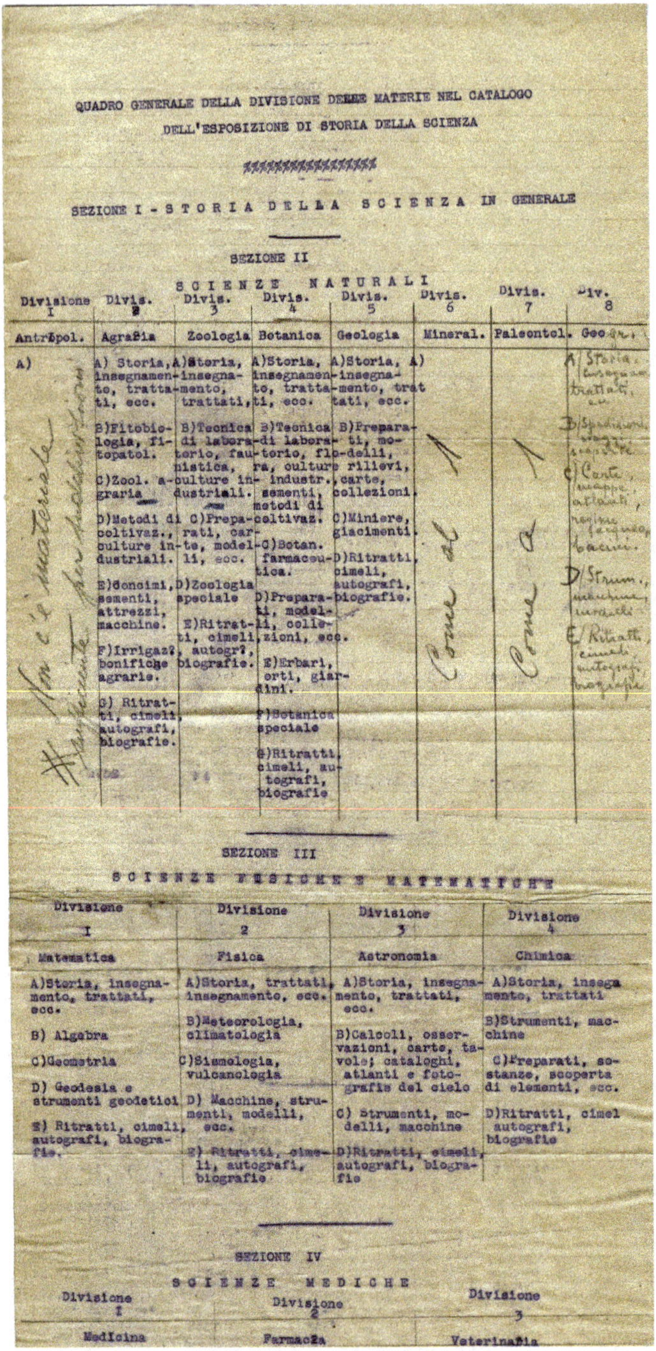

QUADRO GENERALE DELLA DIVISIONE DELLE MATERIE NEL CATALOGO
DELL'ESPOSIZIONE DI STORIA DELLA SCIENZA

SEZIONE I - S T O R I A D E L L A S C I E N Z A IN GENERALE

SEZIONE II

S C I E N Z E N A T U R A L I

Divisione 1	Divis. 2	Divis. 3	Divis. 4	Divis. 5	Divis. 6	Divis. 7	Div. 8
Antropol.	Agraria	Zoologia	Botanica	Geologia	Mineral.	Paleontol.	Geo
A)	A) Storia, insegnamento, trattati, ecc. B) Fitobiologia, fitopatol. C) Zool. agraria D) Metodi di coltivaz., culture industriali. E) Concimi, sementi, attrezzi, macchine. F) Irrigaz? bonifiche agrarie. G) Ritratti, cimeli, autografi, biografie.	A) Storia, insegnamento, trattati, B) Tecnica di laboratorio, faunistica, culture industriali. C) Preparati, carte, modelli, ecc. D) Zoologia speciale E) Ritratti, cimeli, autogr?, biografie.	A) Storia, insegnamento, trattati, ecc. B) Tecnica di laboratorio, flora, culture industr., sementi, metodi di coltivaz. C) Botan. farmaceutica. D) Preparati, modelli, collezioni, ecc. E) Erbari, orti, giardini. F) Botanica speciale G) Ritratti, cimeli, autografi, biografie	A) Storia, insegnamento, trattati, ecc. B) Preparati, modelli, rilievi, carte, collezioni. C) Miniere, giacimenti. D) Ritratti, cimeli, autografi, biografie.	A)		

SEZIONE III

S C I E N Z E F I S I C H E E M A T E M A T I C H E

Divisione 1	Divisione 2	Divisione 3	Divisione 4
Matematica	Fisica	Astronomia	Chimica
A) Storia, insegnamento, trattati, ecc. B) Algebra C) Geometria D) Geodesia e strumenti geodetici E) Ritratti, cimeli, autografi, biografie.	A) Storia, trattati, insegnamento, ecc. B) Meteorologia, climatologia C) Sismologia, vulcanologia D) Macchine, strumenti, modelli, ecc. E) Ritratti, cimeli, autografi, biografie	A) Storia, insegnamento, trattati, ecc. B) Calcoli, osservazioni, carte, tavole; cataloghi, atlanti e fotografie del cielo C) Strumenti, modelli, macchine D) Ritratti, cimeli, autografi, biografie	A) Storia, insegnamento, trattati B) Strumenti, macchine C) Preparati, sostanze, scoperta di elementi, ecc. D) Ritratti, cimeli autografi, biografie

SEZIONE IV

S C I E N Z E M E D I C H E

Divisione 1	Divisione 2	Divisione 3
Medicina	Farmacia	Veterinaria

FIGURE 38 Draft of the general catalogue

FIGURE 39 Andrea Corsini and Maria Luisa Righini Bonelli in the 1950s

PART 2

The Exhibition in Detail

CHAPTER 7

The Exhibition in Detail

The geographic ordering of the Exhibition, as we have seen, was a decision forced on organizers by local committees. Its most immediate effect was to turn the National Exhibition itself into a collection of regional exhibitions loosely held together by the Italian genius myth superimposed on them by the fascist regime. On an historiographical level, this allows us to appreciate the extent of Italian localism and how this localism produced conflicting traditions and narratives that ultimately stood in the way of the developing of a national preservation policy for scientific heritage. Moreover, it also gives us the opportunity to see the coexistence between local narratives and fascist/national narrative in plastic rendition, and the mutualistic interaction between them in its full realization. The decision to let each participant curate its own show, while preventing the Exposition from offering a unitarian view of the development of Italian science, gives us the chance to see different approaches to exhibiting historical scientific collections in action. In general, all the shows featured at the Exposition shared a commendatory purpose as they were meant to celebrate local traditions; yet the strategies deployed to reach this goal significantly differed from one another.

The first and most common approach was the 'hagiographical' one adopted by small and/or peripheral cities, and largely focused on individuals. In it, artefacts were treated as relics and exhibits as reliquaries; hence written material, portraits and memorabilia such as medals, plates, pieces of furniture and clothes were given pre-eminence over instruments and machines. The best example of this approach is the exhibition of Bergamo, managed by the director of Bergamo's Civic Library, Giuseppe Locatelli, and mostly devoted to the Bergamese explorer Giacomo Costantino Beltrami: not only Beltrami's writings and maps, but also his trousers and aprons were on display.

The second extensively adopted approach was the one focusing on institutions rather than individuals. Though celebratory in tone just like the hagiographical approach, it emphasized publications, portraits, scientific instruments, and laboratory apparatuses as attestations of the research activity carried out in the institutions they belonged to. The case of Bologna is particularly illustrative of this approach. The committee established in 1927 featured some of the leading scientific personalities of the town, representatives from city museums and archives, University professors, and delegates of the City Council,

 | DOI:10.1163/9789004297418_009

the Provincial Council for Economy, and Bologna Savings Bank. The setup of the room was managed and curated by an executive board consisting of mathematician Ettore Bortolotti, professor of Geodesy and Topography Paolo Dore, archaeologist Pericle Ducati, professor of Anatomy Ercole Giacomini, and agronomist Dino Zucchini. At that time, Bologna was an active center of research on historical scientific collections: the University had already staged an exhibition of antique science books in 1922 and at the time of the Florentine Exposition an exposition in honour of Luigi Ferdinando Marsili, founder of the Bolognese Institute of Science, was on the way to be held on the second centenary of his death in 1930. The central theme of the room was the celebration of the University – the oldest in the world – which in 1928 had celebrated the 840th anniversary of its foundation. "The main concept followed by the Bolognese Committee was to celebrate the scientists who worked at the University," an official statement of the committee declared in 1929, "and to play up, albeit sketchily, the continuity of Bologna's contribution to observative, experimental, and speculative science."[1] The idea was to highlight the continuity of a tradition that, starting from Gerolamo Manfredi and Ulisse Aldrovandi, and renewed by figures like Raffaele Bombelli, Marcello Malpighi, Giovanni Domenico Cassini, Luigi Ferdinando Marsili, and Luigi Galvani, culminated in Augusto Righi and, above all, Guglielmo Marconi, the "prototype of Fascist scientist," whose instruments and inventions, orderly disposed in a glass case, testified to the renewed strength of the University of Bologna.

If the institutional approach adopted by cities like Bologna retained some element of hagiography, a more decisive step towards the displaying of historical scientific artefacts as 'cultural' objects was taken in the exhibits of Florence, Milan, and Rome. In these shows, artefacts were presented not as mute evidences of the scientific life of individuals or institutions, but as essential elements of cultural identity: by celebrating a local tradition, artefacts were meant to show the role and ambitions of the community that produced them. From this perspective, recreations, replicas and scale models of original artefacts acquired a fundamental role. As the main aim of the objects on display was to convey the idea of an active cultural community, it was of relative importance whether they were the original ones. Within the hagiographical approach, there was no point in showing replicas of scientific instruments and machines, for in such perspective the value of the objects consists precisely in their being the very ones belonged to the scientists they are meant to celebrate; but if scientific artefacts are considered as part of a cultural identity, then building replicas of them to be showed to the public is not only allowed

1 "La sala di Bologna alla Mostra della Scienza," *Unità Cattolica*, 7 August 1929.

but even recommendable. What Corsini hoped for, however, was far away to be realized: an 'ecology' of historical scientific institutions; a network of local museums that exchanged information about their collections between them, and a National Museum (the Florentine one), charged with the task of maintaining an up-to-date catalogue of objects. It was only in 1993 that a work team of the now-called Institute and Museum of History of Science in Florence (the heir of the original National Museum), was able to fulfil, at least in part, Corsini's project with the creation of a cataloguing scheme for scientific objects, the STS (*Scheda per la catalogazione degli strumenti scientifici*), which is now a cataloguing standard recognized by the Italian Ministry of Tourism and Cultural Heritage and Activities.

1 Ground Floor I

1.1 *Tuscany*

The Tuscany room, located right after the entrance, was one the Palace of Expositions's biggest rooms. It hosted the exhibits of Pisa, Arezzo, Pistoia, Larderello, Siena, Volterra, Prato, Lucca, Livorno, and Empoli, plus a thematic exhibition on Leonardo da Vinci. In total, there were 582 items on display.

Pisa's participation in the First National Exhibition of History of Science was coordinated by the University and involved some of the city's leading scientific institutions: the School of Applied Engineering (*Scuola di Applicazione per gli Ingegneri*), the *Scuola Normale Superiore*, the Applied Physics Laboratory, and the Department of Zoology. In total, these institutes presented 91 items, more than half of which belonged to the School of Applied Engineering. Founded in 1913 and directed by Giovanni Quaglia, the School owned relics of remarkable historic value like Pieter van Musschenbroek's air-pump and Antonio Pacinotti's drawings, manuscripts, and electromagnetic devices.[2] The latter relics – held by the Institute of Engineering Physics, annexed to the School – were on public display for the first time ever. They had been located and indexed following the reorganization of the Institute's archives promoted by the new director Giovanni Polvani in 1927 and would later become part of the "Pacinotti Museum" established in 1930 and inaugurated in 1934.[3] The *Scuola Normale*

2 See Giuseppe Polvani, *Le cose storicamente più notevoli possedute dall'Istituto di fisica tecnica della R. Scuola d'Ingegneria di Pisa* (Pisa: Pubblicazioni della R. Scuola d'Ingegneria di Pisa), n. 147, September 1930; and Giuseppe Polvani, *Antonio Pacinotti: la vita e l'opera* (Pisa: Lischi, 1934).

3 Claudio Luperini and Tiziana Paladini, *Ho costruito il seguente apparecchietto. Antonio Pacinotti, manoscritti e strumenti* (Pisa: ETS, 2007), 1–14.

Superiore, led by Giovanni Gentile, celebrated its past by means of manuscripts and portraits of its directors in order to support its claim for administrative autonomy from the University of Pisa, this being one of Gentile's main concern as director.[4] The Pisan exhibits were enriched by a collection of portraits sent by the University, the complete series of the *Atti della Società Toscana di Scienze Naturali*, two 18th Century celestial globes, a number of books, and several photographs.

Among the Tuscan cities, Siena was the one which contributed the largest number of objects (177), even though they were mostly books. The participation of Siena had been promoted by Corsini's friend and colleague Domenico Barduzzi, president of the *Accademia dei Fisiocritici* (Academy of Physiocritics). The Academy, founded in 1691 by physician Pirro Maria Gabrielli, had begun collecting scientific artifacts in the XVIII century and by 1929 held a sizable collection. After Barduzzi's death in February 1929, Ludovico Tommasi succeeded him in the position and took care of the exhibit. This was centered around the figures of Gabrielli, Paolo Mascagni, and engineer Giovanni Caselli. As said, the majority of the objects on display were books; still, there were a few portraits, manuscripts, and photographs too. The initiative received financial support from the *Monte dei Paschi* Bank.[5] There were three other exhibits mostly consisting of books: those of Pistoia, Volterra, and Empoli. In Pistoia, a local committee presided by physician and historian of medicine Alberto Chiappelli was established in 1927.[6] Since 1907, Chiappelli had been publishing papers on the history of medicine, focusing in particular on the 17th century Pistoiese medical school. Surveying the archives of the *Ospedale del Ceppo*, Chiappelli managed to set up an exhibit that celebrated the Pistoiese contribution to the history of medicine. He himself contributed with books and manuscripts from his private library. Centered on the history of medicine was also the exhibit set up by the City Council of Empoli, which featured works by Empolese physicians Vincenzo Chiarugi and Giuseppe Del Papa. Finally, Maurizio Cavallini, a priest and historian who was then busy working on the establishment of the *Museo di Arte Sacra* of Volterra (inaugurated in 1932), managed the setup of the Volterra exhibit. It consisted of 55 items and included a collection of scientific publications, a working model of Volta's battery, compasses, nautical charts, and theodolites.[7]

---

4 See Tina Tomasi and Nella Sistoli Paoli, *La Scuola Normale Superiore di Pisa dal 1813 al 1945* (Pisa: ETS, 1990), 171–214.

5 "Per l'Esposizione di Firenze," *Unità Cattolica*, May 21, 1929.

6 Alfredo Chiti, "La prima Esposizione Nazionale di Storia della Scienza (Firenze, 1929 VII). Ricordo del comitato provinciale pistoiese," *L'Azione*, May 11, 1929.

7 "Volterra alla I Esposizione di Storia della Scienza," *il Corazziere*, May 19, 1929.

In Lucca, a local committee was established in 1927. It was presided by Eugenio Lazzareschi, director of the State Archive of Lucca, and featured representatives from the Government's Library of Lucca and the "Niccolò Machiavelli" High School. The latter institution owned a huge collection (almost 400 items) of physics equipment dating back to its foundation in 1789, and samples from this collection were sent to Florence.[8] Another local committee was established in Livorno in the same year 1927, and was presided by Giulio Cappadoro, president of the Naval Academy. However, since the Naval Academy itself staged a separate exhibit in the Italian Navy Pavillion, the Livornese exhibit mounted in the Tuscan room consisted of books and manuscripts from the *Biblioteca Labronica* only.[9] The city of Prato presented 38 items grouped in two sections: the first one, consisting of manuscripts, terrestrial and celestial globes, armillary spheres, and samples from Andrea Guazzalotti's medal collection, came from the *Biblioteca Roncioniana*; the second one, coming from the newly-discovered Datini archive, consisted of cloth samples, drawings of looms, and instructions for preparing colours.[10] The setup was curated by Ovidio Ballerini, librarian at Roncioniana and archivist of the Datini archive, with the help of Ruggero Nuti, chief archivist of the City of Prato, under the supervision of Angiolo Badiani, president of the Florence Province and Superintendent for Fine Arts in Prato.[11] The realization of the Larderello exhibit was prompted by Piero Ginori Conti in order to show the history and development of the *Società Boracifera*. It was curated by Riccardo Grassina and consisted of steam engines, pumps, photographs and scale models of the factory; publications, portraits, and medals.[12]

The very center of the Tuscan room was the thematic exhibition on Leonardo da Vinci. It was intended to celebrate the highest point reached by the Italian and Tuscan genius and was devised and curated by Corsini himself. The exhibition revolved around Leonardo's fascination with flight and featured a bust of Leonardo, copies of drawings and figures from *Codex Atlanticus* and *Codex on the Flight of Birds*, and, above all, a collection of models built for the occasion by Raffaele Giacomelli, one of the leading Italian experts on Leonardo's theories of flight. Corsini contacted Giacomelli in 1928 and asked him to work

8 "Una Esposizione nazionale di Storia della Scienza a Firenze," *il Popolo toscano*, August 1, 1927.

9 "Una Esposizione Nazionale a Firenze di storia della scienza," *Gazzetta livornese*, August 6, 1927.

10 Francesco Datini (1335–1410), a merchant from Prato, left an extensive archive discovered only in the 19th Century.

11 "La partecipazione di Prato alla mostra di Firenze," *l'Avvenire d'Italia*, May 10, 1929.

12 Italo Del Giudice, "Le industrie italiane alla Mostra delle Scienze," *Unità Cattolica*, August 25, 1929.

on a philologically sound reconstruction of Leonardo's machines. Working on Leonardo's *Codex Atlanticus, Codex B*, and *Codex on the Flight of Birds*, Giacomelli sketched a set of directions which were transmitted to Giuseppe Schneider, a designer at *Stabilimento Costruzioni Aeronautiche* (Aeronautical Constructions Company) in Rome. Some of the models were built by Mario Bucci and Vasco Menici, craftsmen of the "Leonardo da Vinci" Polytechnic Institute in Florence, and others by engineers at the *Stabilimento*.[13] The collection included four flying machines, two wing models, the parachute and the aerial screw, a lathe, and a couple of devices illustrating how the wings were supposed to twist. The exhibition was completed by a brass instrument, built by Neapolitan professor Roberto Marcolongo, used by Leonardo to solve Alhazen's problem.

In his explication of the work done, Giacomelli insisted on the fact that what he actually wanted was to present to the public a reconstruction as accurate as possible of Leonardo's flying machines. Unlike Paul Garber, the American historian of aviation and later director of the National Air Museum of Washington, D.C., who in the same year 1929 had presented his reconstruction of Leonardo's flying machine, Giacomelli made it a point to realize a "plain and thorough reproduction" of Leonardo's drawings:

> The public [...] generally believes that Leonardo's failure in achieving human flight depended on his lacking of a proper engine, and therefore thinks that it is now possible to build Leonardo's flying machine following the instructions written down in his manuscripts. Still, as a matter of fact, in Leonardo's manuscripts one can find no drawing representing a full and complete flying machine, but only sketches of different machines and drawings of specific devices. [...]
>
> Mr. Paul Garber's model differs from ours because his is not a plain and thorough reproduction of a Leonardo's drawing (the main body of the machine together with the veining of the wing) but it is a full flying machine obtained by combining the main body of a machine as it is sketched in a drawing with a couple of wings from another drawing.[14]

---

13 Antonio Poggi, "Le opere immortali di Leonardo e Galileo alla Mostra Nazionale di Storia delle Scienze," *la Nazione*, May 7, 1929; Dino Brogi, "La macchina volante di Leonardo ricostruita alla Mostra della Scienza," *il Giornale d'Italia*, June 26, 1929; Italo Del Giudice, "La ricostruzione della macchina volante di Leonardo," *Unità Cattolica*, July 28, 1929.

14 Raffaele Giacomelli, "I modelli delle macchine volanti di Leonardo da Vinci," *l'Ingegnere*, v (2), 1931.

After the Exhibition closed, the models built in Florence remained in the holdings of the Museum of History of Science, whilst those built in Rome were donated to the Science Museum of London. Both were on display at the Milan Aeronautical Exhibition in 1934, and served as guidelines for the realization of the models on display at the great exhibition on "Leonardo and the Italian Inventions" held in Milan in 1939 (on that occasion, Giacomelli oversaw the setup of the "flight room").

1.2 *Florence*

Florence represented both physically and symbolically the very center of the Exhibition. It was celebrated as "the cradle of modern science,"[15] "the spiritual center of Italy,"[16] and "the capital of Italian Genius."[17] Being the Queen City of the Exhibition, it occupied the "place of honour," the inner central room, in order to physically represent its being the origin and truest essence of the Italian scientific tradition. The layout was curated by Corsini, Garbasso, and Pagnini, and focused on portraying and celebrating the Galilean tradition in the figures of Galileo himself, the academicians of Cimento, and their supposed followers. To stage it, the curators took advantage of Galileo's tribune's and *Museo degli strumenti antichi*'s holdings, but also borrowed objects from the Uffizi Gallery and from private collectors like the count David Augusto Costantini. In total, there were about 250 objects on display.

The hall layout was guided by esthetic and functional criteria more than by historical or educational priorities. The largest instruments – such as the Gonella, Amici, and Campani telescopes and the machines for performing physical experiments – were placed at the center. The two Coronelli globes flanked the entrance. At the far side of the room stood Peter Leopold's chemistry cabinet, and all along the walls instruments and machines coming from the *Museo degli strumenti antichi* were lined. Hanging on the same walls, a collection of portraits of Florentine scientists and Cimento Academicians borrowed from the Uffizi Gallery. On the overall, the figure of Galileo prevailed. Galilean relics were scattered throughout the room so as to signify that every line of research being typified in the items on display had its own origin in Galileo's work. On the left of the room, Galileo's telescopes were on display in a showcase; on the opposite side, seated in a case, Galileo's compass and armed

15 "Le glorie di Firenze alla Mostra della scienza," *Unità Cattolica*, September 20, 1929.
16 Giorgio Abetti, "Il centro spirituale d'Italia," *la Nazione*, December 1, 1929.
17 Alessandro Pavolini, "Vin buono," *il Bargello*, June 9, 1929.

lodestone; in the center, "Galileo's astrolabe";[18] on the walls, Galileo's drawings of the Moon, the Medicean Planets, and Sunspots; in a huge bookcase, the complete *Edizione Nazionale* of Galileo's writings; and finally, enclosed in a shrine, the lion's-foot-shaped feet of Galileo's bed.

Though openly celebratory in tone, this setup was not as religiously tinged as the Galileo's Tribune's one. Galileo was celebrated as the father of experimental method and not as a Christian saint. Curators thought it better not to put on display Galileo's middle finger, for example, and even official press releases insisted more on the "experimental tradition" supposedly originated from Galileo's work than on his life and personality. For these reasons, curators chose to put emphasis on instruments, machines, and apparatuses, rather than on books and manuscripts. In a sense, the setup was meant to suggest the identity between the Galilean tradition and the Italian scientific tradition by means of the idea of experimentation developed by Galileo: as Abetti put it, "the whole Exhibition radiates from Galileo."[19]

Curators excluded from this room everything not directly related to the Galilean experimental tradition: anatomical waxes, herbals, and surgical instruments were all put on display in another room located in the basement. This decision was taken not only for obvious reasons of space, but also to start of a consistent collection of objects for the planned new Museum. In 1928, Garbasso had already sought the consent of Enrico Burci, Rector of the University of Florence, for the deposit of a portion of the physical and astronomical instruments of the *Museo degli Strumenti Antichi* in a new Museum of Physics. Modeled on the Deutsches Museum, the new institution would be situated in the garden of the Reale Società Toscana di Orticoltura purchased by the Municipality of Florence. Despite the Rector's consent, Garbasso failed to raise sufficient funds for the project, and so decided to fully endorse the plan promoted by Corsini and Ginori Conti. The 1929 Exposition thus provided the occasion to select the items and test a display option involving a broad separation between physics, astronomy, and optics on the one hand, and medical and natural sciences on the other. This division was eventually adopted in the initial layout of the new Museum inaugurated in 1930.[20]

18 With this name it is known an astrolabe used by Galileo for his astronomical calculation. Once preserved in the Uffizi Gallery, the instrument is now on display at the Galileo Museum of Florence.

19 Giorgio Abetti, "L'edizione nazionale delle Opere di Galileo e la Mostra delle scienze di Firenze," *l'Universo*, 10, 10 (1929), 1007–1014: 1014.

20 See chapters 2 and 5.

1.3 *Rome*

The Roman collections were housed in a room next to the Tuscany room and large enough to accommodate almost 300 objects mostly coming from the Copernican Museum and the Astronomical Observatory of Monte Mario. The local committee that managed the participation at the Exhibition was established in 1927. It was presided by Federico Millosevich (Rector of "La Sapienza" University of Rome) and featured a large number of personalities from the political, cultural, scientific, and military world.[21] However, the task of selecting, sorting, and sending the items to Florence was actually performed by an *ad hoc* committee consisting of Francesco Boncompagni Ludovisi (as President), Millosevich (as Director), and Giuseppe Montalenti (as Executive Secretary). The layout of the room was curated by Montalenti.[22]

The curator chose to focus on astronomy, medicine, and technology. The Copernican Museum's collections – which featured, among other things, 31 telescopes, 13 astrolabes, 12 armillary spheres, 18 celestial globes, 26 microscopes, and 41 sundials – were put on display side by side in three showcases; next to them, the Lincei Academy's collection, which included manuscripts, portraits, and medals celebrating the institution's history and members.[23] The Academy was then going through a big crisis: being looked suspiciously by the fascist regime, in 1933 it would be put under the administration of an external commissioner and then, in 1939, incorporated into the regime-controlled *Accademia d'Italia*.[24] The medicine section consisted of anatomical waxes coming from the *Ospedali Riuniti* and samples from Giovanni Carbonelli's collection. The latter had been recently donated to the City Council of Rome, and featured microscopes, surgical instruments, anatomical models, portraits, books, and letters.

A significant portion of the room was devoted to the celebration of Imperial Rome, in a fashion that was characteristic of fascist colonial propaganda. In 1928, the Regime had acknowledged the *Istituto Coloniale Fascista* (Fascist Colonial Institute) as the only institution officially authorized to pursue

---

21 "Una esposizione nazionale di storia della scienza a Firenze," *il piccolo Giornale d'Italia*, August 6, 1927.

22 "La partecipazione di Roma alla Mostra di storia delle scienze. Dall'arte costruttiva e militare dei Romani alle Mostre dell'Accademia dei Lincei e dell'Osservatorio astronomico del Campidoglio," *la Tribuna*, May 12, 1929.

23 Giuseppe Gabrieli, "La romana Accademia Lincea nella Esposizione fiorentina di storia della scienza," *Capitolum*, June 1929, 302–315: 313.

24 See Raffaello Morghen, *L'Accademia Nazionale dei Lincei nel CCCLXVIII anno dalla sua fondazione, nella vita e nella cultura dell'Italia unita (1871–1971)*, (Roma: Accademia Nazionale dei Lincei, 1972), 61–66.

colonial propaganda and entrusted it with the task of shaping "colonial awareness" among the Italian population. The Institute published books and periodicals, managed school programs, promoted local and national initiatives, and participated in major events such as exhibitions and trade fairs.[25] At the Florence exhibition, a great emphasis was put on "Rome's civilizing mission": an entire wall was covered with plaster casts of Trajan's column representing siege engines, military bridges, ships, cranes, aqueducts, sewerage systems, etc., so as to highlight the level of sophistication reached by Roman technology and how the military campaigns contributed to spread it around the world – a not-too-hidden allusion to Mussolini's unconcealed wish to rebuild the "Italian Empire." In the centre of the room stood a sculpture of the Capitoline Wolf, a symbol of the "Third Rome" Mussolini yearned for: "ample, neat, powerful, just like it was back in the times of Augustus."[26]

1.4 *Bologna*

The committee established in Bologna in 1927 was presided by Corsini's friend and colleague Domenico Majocchi (after Majocchi's death on March 7, 1929, zoologist Alessandro Ghigi succeeded him in the office). It featured some of the leading scientific personalities of the town, representatives from city museums and archives, University professors, and delegates of the City Council, the Provincial Council for Economy, and Bologna Savings Bank.[27] The setup of the room was managed and curated by an executive board consisting of mathematician Ettore Bortolotti, professor of Geodesy and Topography Paolo Dore, archaeologist Pericle Ducati, professor of Anatomy Ercole Giacomini, and agronomist Dino Zucchini.[28]

At that time, Bologna was an active center of research on and protection of historical scientific collections: there were projects to valorize the Aldrovandi Museum, the *Museum Diluvianum*, and the collections of the Astronomical Observatory directed by Guido Horn d'Arturo.[29] The University had already staged an exhibition of antique science books in 1922 and, after the Florence exhibition closed, it would organize another in honour of Luigi Ferdinando

25 On this argument, see Mariella Cagnetta, *Antichisti e Impero fascista* (Bari: Dedalo, 1979).

26 Benito Mussolini, *La nuova Roma*, in Benito Mussolini, *Opera Omnia*, vol. XXII (Firenze: La Fenice, 1957), p. 48.

27 "La partecipazione di Bologna all'Esposizione di storia della scienza," *Resto del Carlino*, August 9, 1927.

28 "Il contributo di Bologna alla mostra di storia delle scienze," *Resto del Carlino*, February 9, 1930.

29 See chapter 1.

Marsili, founder of the Bolognese Institute of Science, which was held on the second centenary of his death in 1930.

The central theme of the room was the celebration of the University – the oldest in the world – that in 1928 had celebrated the 840th anniversary of its foundation. "The main concept followed by the Bolognese Committee was to celebrate the high scientists who worked at the University," an official statement of the committee declared in 1929, "and to play up, albeit sketchily, the continuity of Bologna's contribution to observative, experimental, and speculative science."[30] Hanging on the walls there were portraits of the University's most renowned scientists, samples of whose works were on display in showcases running longitudinally through the room and in glass and book cases lined along its sides and framed by busts and other memorabilia.

The idea was to highlight the continuity of a tradition which, starting from Gerolamo Manfredi and Ulisse Aldrovandi, renewed by figures like Raffaele Bombelli, Marcello Malpighi, Giovanni Domenico Cassini, Luigi Ferdinando Marsili, and Luigi Galvani, culminated in Augusto Righi and, above all, Guglielmo Marconi, the "prototype of Fascist scientist," whose instruments and inventions, orderly disposed in a glass case, testified to the renewed strength of the Italian (and Bolognese) Genius. In total, there were 235 objects on display: 139 manuscripts and books, 34 portraits, 6 busts, 5 wood engravings, 1 miniature, 23 instruments, 16 preparations, 6 photographs, 2 portolan charts, 3 paintings, and 2 antique books on Italian and Neapolitan horse breeds.[31]

2 Ground Floor II

2.1 *Lombardy*

The Lombardy room housed, "cuddled in a brotherly way,"[32] the exhibits of Bergamo, Brescia, and Pavia. The main promoter of the participation of Bergamo had been the director of the Civic Library of Bergamo, Monsignor Giuseppe Locatelli, who selected about 200 objects related to Bergamese scientists. Most of them were related to Giacomo Costantino Beltrami, the Bergamese explorer who discovered the sources of the Mississippi river in 1823. Beltrami's archive of correspondence and relics were housed at the Civic

30 "La sala di Bologna alla Mostra della Scienza," *Unità Cattolica*, August 7, 1929.

31 For further details see Marco Beretta, *L'Alma Mater all'Esposizione Nazionale di Storia della Scienza di Firenze (1929)*, in *Una scienza bolognese? Figure e percorsi della storiografia della scienza a Bologna dal Risorgimento a oggi*, edited by Annarita Angelini, Marco Beretta, and Giuseppe Olmi (Bologna: Bononia University Press, 2015), 263–277.

32 *Guida della prima esposizione nazionale di storia della scienza*, cit., 45.

Library of Bergamo and at the Natural History Museum, respectively, and their indexing and cataloguing were still at an initial phase. The collection on display in Florence included Beltrami's works and correspondence, portraits, maps, instruments, tools, and even pieces of clothing and furniture.[33] In Brescia, a local committee created in 1927 within the University and presided by director of Brescia State Archive Enrico Boggiano chose and sent to Florence 166 objects. The exhibit featured the complete collection of Niccolò Tartaglia's and Francesco Lana de Terzi's works, and a small collection of instruments consisting of astrolabes, compasses, and solar clocks. The research work was carried out by Ottavio Trainini, Angelo Ferretti Torricelli, and Francesco Massardi, who managed to get involved institutions like the *Biblioteca Queriniana* and the Civic Museum of Brescia.[34]

If the exhibits of Bergamo and Brescia celebrated the figures of Beltrami, Tartaglia, and Francesco Lana de Terzi mostly by exhibiting their written works and correspondence, the exhibit of Pavia, on the contrary, celebrated Alessandro Volta, Paolo Mantegazza, Luigi Valentino Brugnatelli, and Antonio Scarpa by putting on display their instruments. The curators Guido Sala and Antonio Pensa benefited from the rich collection belonging to the Physics Cabinet of the University of Pavia,[35] which, together with the anatomical collections gathered by Antonio Scarpa, were then undergoing major restoration work that would eventually lead to the constitution of the Museum for the History of the University of Pavia, inaugurated in 1932, on the first centenary of Scarpa's death.[36] The exhibit was spread over two walls. A showcase on one side of the room housed Paolo Mantegazza's instruments, Brugnatelli's electroplates, and a small collection of 15th Century medical texts. On the opposite side, instruments, preparations, and texts by Antonio Scarpa and his students and followers. Alessandro Volta's surviving relics were placed in a showcase in the center of the room.

---

33 "Bergamo alla Mostra nazionale di storia della scienza a Firenze," *l'Eco di Bergamo*, May 8, 1929.

34 "Una esposizione nazionale di storia della scienza a Firenze," *il Popolo di Brescia*, August 10, 1927. For detailed information about the participation of Brescia, see Donatello Romano Dall'Asta, *La partecipazione di Brescia alla prima Esposizione nazionale di storia della scienza a Firenze nel 1929*, in *L'Ateneo di Brescia e la storia della scienza* (Brescia: f.lli Geroldi, 1985–1988), 25–52.

35 See ch. 1.

36 See Lucio Fregonese, "Il Museo per la storia dell'Università di Pavia: storia, patrimonio e nuovi allestimenti," in *Annali di storia delle Università italiane*, vol. 5 (2001).

2.2 *Milan*

Milan, the self-styled 'moral capital of Italy', exhibited, in a relatively small room, 387 objects, most of them printed books and pamphlets. The local committee was presided by Francesco Somaini, the member of the Parliament who had already promoted and financed the construction of the *Tempio Voltiano* in Como. The layout, besides celebrating scientists who were born or had worked in Milan – like Bonaventura Cavalieri, Ruggero Boscovich, Paolo Frisi, and Giovanni Schiaparelli – was intended as a celebration of entrepreneurship: the centre of the room was occupied by a *Magrini* electric engine, one rotor from the first turbine built by Riva-Monneret company, and a model of the first locomotive engineered in a workshop near Porta Tosa in 1853. The City Council of Milan staged two exhibits on aeronautics and navigation: the exhibit on aeronautics was staged thanks to the collection donated to the City by Achille Bertarelli and consisted of documents and materials on aeronautical experiments performed by Francesco Zambeccari and Vincenzo Lunardi; the exhibit on navigation, on the other side, featured pictures of ships coming from the "Naval Museum" created by Filippo Camperio and Federico Jarach, and donated to the City of Milan in 1922. Other institutions and privates staged their own exhibits. The Brera Observatory directed by Emilio Bianchi sent a number of publications documenting its research activities since 1777; Tommaso Gnoli, director of the Braidense Library, curated an exhibit on the history of geography which featured texts and objects from the Braidense and the Ambrosiana libraries. Finally, Carlo Calzecchi-Onesti, superintendent of Fine Arts and son of physicist and inventor Temistocle, provided instruments, relics, and memorabilia regarding or belonged to his father.

On the overall, the Milan room was meant to suggest the idea of a dynamic industrial city. Its main theme – transportations – conveyed the idea of an uninterrupted technological development and continuity since the International Exhibition of 1906.[37] At the same time, the stress put on industry and technology signalled the role that Milan was trying to secure in those years as the most adequate seat for a National Museum of Science, Industry, and Technology. In 1927, engineer and entrepreneur Guido Ucelli, having received support and encouragement from Giuseppe Belluzzo, had begun rallying Milanese cultural and political institutions towards this end.[38] Milan was then redefining its identity as the preeminent industrial and scientific center of Italy, more or less like Florence was rushing to be acknowledged as its Cultural capital. Throughout the 1920s, significant initiatives like the foundation of the University of Milan

---

37 See ch. 1.

38 See ch. 6.

(1924), the enlargement and improvement of the Brera Observatory (1926), the inauguration of the National Institute for the Cure from Cancer (1928), and the completion of Italy's largest planetarium (1930) vouched for Milan's renewed push on science and technology. Thus, when Ucelli's project was announced, reactions were enthusiastic and not so much different, in their localistic celebratory tone, from those uttered in Florence at the news of a possible Museum of History of Science. "We must not tolerate any discussion about the location," a Milanese newspaper remarked. "Only Milan, as the mother and queen of the Italian industry, is worthy of the duty and honour of housing such a great institution."[39]

2.3 *Turin*

Milan's claim of being the leading Italian industrial and scientific city was challenged by Turin, the city which, in its three rooms, housed motor engines and even a complete car engineered by FIAT, the preeminent Italian machine company. The Piedmontese participation at the Exhibition was coordinated by a committee presided by Carlo Baravalle, Achille Bologna, and Stefano Bricarelli, and featuring representatives from the University of Turin, the Polytechnic, the State Archive, the National Library, the Engineering School, the Astronomical Observatory, the Society of Industrial Chemistry, the Museum of Zoology and Comparative Anatomy, and the Agricultural Academy.[40] The exhibit was curated by Pietro Gorgolini, a fascist activist and director of the newspaper *il Nazionale*, and by Carlo Federico Bonini, librarian at the Polytechnic.[41]

In general, the exhibit focused on Turin's contribution to the development of science and on priority claims. According to *il Nazionale*, the Turin exhibit succeeded in "demonstrating" the 'Italianity' of inventions and discoveries such as the typewriter, devised by Giuseppe Ravizza, and the nitroglycerin, first synthesized by Ascanio Sobrero. Furthermore, the exhibit commemorated the II Congress of the Italian Scientists held in Turin in 1840 and, more generally, reminded the visitor that, throughout the 19th Century, Turin had been the Italian city which contributed the most to the development of science and technology. "It is truly a great collection of authentic Italian glories," *il Nazionale* gloated, "and reaffirming them will admonish the world and encourage the Italian spirit."[42]

---

39 Metron, "Un Museo Industriale Italiano," *Corriere della sera*, September 15, 1931.

40 "Una Esposizione nazionale di storia della scienza," *Momento*, August 6, 1927.

41 "Torino alla mostra della scienza," *il Nazionale*, October 5, 1929.

42 "Torino alla mostra della scienza," *il Nazionale*, October 5, 1929.

A very special place within the Turinese exhibit was assigned to the figure of Galileo Ferraris. He was celebrated as the first scientist who studied rotating magnetic fields, and some of his instruments and devices were on display in a showcase. Ferraris was born in Livorno Piemonte in 1847; in 1925, his family donated the building where he was born to the city, which on the occasion changed its very name to "Livorno Ferraris" and set up a museum in the scientist's house. The museum was known as "the Ferraris Sacrarium" and was inaugurated in 1931, following the success of the Turinese exhibit in Florence.[43]

2.4 *Genoa*

The Genoa exhibit consisted of 97 objects divided into iconography (32 items), instruments and machines (30 items), books and texts (21 items), and cartography (14 items). The institutes that contributed the largest number of objects were the University of Genoa (26 items), the Hydrographic Institute (35 items), and the Ansaldo Naval Shipyards (16 items). The participation was coordinated a local committee created by the University of Genoa and featuring, among others, mathematician and historian Gino Loria and representatives from the City of Genoa, the University, the State Archive, the "Berio" Civic Library, and the Writers' Union of Liguria.[44]

In 1929, Genoa was experiencing a significant growth which in 1930 would eventually lead to the official constitution of the so-called "Big Genoa" (*la grande Genova*). Launched in 1926 with the City's annexation of 19 smaller peripheral municipalities like Voltri and S. Ilario, it was a project strongly pushed by Mussolini himself in order to balance out Genoa's effective size with its foreign trade volume. It was officially achieved in 1930 with the approbation of a new town planning scheme which would guide all major initiatives of urban renewal throughout the 1930s.[45] Thus, in accord with this project, the exhibit focused on naval science and hydrography. The Ansaldo Naval Shipyards, which was then entering a profound crisis and would bankrupt in 1932, presented a collection of 16 photographic albums showing pictures of ships built during the 19th Century, while the Hydrographic Institute sent a collections of navigation instruments and devices for the study of sea currents. The University, for its part, exhibited portraits of Genoese scientists and ancient medical texts.

---

43 "L'inaugurazione del 'Sacrario Ferraris'," *l'Elettrotecnica*, vol. XVIII, nn. 23–24 (1931), 582–584.

44 "Genova all'Esposizione di Storia della Scienza," *il Secolo XIX*, October 21, 1928.

45 See Giulio Ernesti, *La grande città nel ventennio fascista*, in *La grande Genova 1926–2006. Atti del Convegno di Studi, Genova, 28–30 novembre 2006* edited by Elisabetta Arioti, Luigi Canepa, and Raffaella Ponte (Genova, 2008), 117–144.

Genoa shared the room with Ferrara, Pallanza, and Cremona. These three cities exhibited very little, so it was judged more convenient to house them in the Genoa room. Ferrara contributed a fifty-volume herbal compiled by Antonio Campana, director of the Botanical Garden of Ferrara from 1804 until his death in 1832. The herbal arrived in Florence together with two in-folio books on the reclamation of the area between the Po and Tartaro rivers (1608), courtesy of Eugenio Baroni, current director of the Botanical Garden. Pallanza and Cremona exhibited only an etching of the first sailing steamship (the 'Verbano') on the *Lago Maggiore*, while the Cremona showcase consisted of an armillary sphere, a microscope, and three sundials property of the "Daniele Manin" High School. The Cremonese collections, a local newspaper explained, were actually very rich; still, due to practical inconveniences, it had been impossible to send even a part of them to Florence.[46]

3 Ground Floor III

3.1 *Veneto*

The room called "Veneto" housed the exhibits staged by the cities of Padua, Verona, Venice, Treviso, Vicenza, Belluno, and Bassano. The regional committee established in 1927 and headquartered in Verona was presided by agronomist and politician Luigi Messedaglia. It featured, among others, botanists Achille Forti and Caro Massalongo (who died in 1928); historian Antonio Avena (director of the Civic Museums of Verona); Vittorio Dal Nero and Giuseppe Stegagno, (director and curator of the Natural History Museum of Verona, respectively); and physician and naturalist Enrico Cartolari.[47]

The exhibit of Padua was staged by the University, the Astronomical Observatory, and the Episcopal Seminary of Padua. An important section of the exhibit was devoted to engineer Enrico Bernardi, died in 1919, whose works were on display thanks to the collaboration of his heirs. There were engines, vehicles, and instruments that in 1941, on the centenary of Bernardi's birth, would flow into the new Museum of Machines "Enrico Bernardi" established within the University of Padua. Besides Bernardi's relics, Padua also presented several instruments coming from the Astronomical Observatory and a collection of medical and pharmaceutical books sent by the University. The Episcopal Seminary of Padua presented a selection of scientific books published by the Seminary starting from 1684, which showed the institution's interest in

46 "Cremona alla I Esposizione Nazionale di storia della scienza," *Cremona*, January 1930.

47 "Una Esposizione Nazionale di Storia della Scienza a Firenze," *l'Arena*, August 7, 1927.

scientific matters.[48] Verona celebrated *Monsignor* Giuseppe Zamboni, who figured as precursor to Alessandro Volta in the invention of dry battery. The exhibit was curated by Gino Sandri, archivist of the City of Verona. The objects on display – several kinds of batteries, an electroscope, a pendulum, and a 'tightrope walker' used for the study of the center of gravity – belonged to the "Scipione Maffei" High School, the institution where Zamboni taught. The setup was completed by a collection of scientific books coming from the Civic Library of Verona and photographs of the Natural History Museum's collections. Venice was represented by the Meteorological-Geodynamic Observatory of the Episcopal Seminary of Venezia directed by Francesco Saverio Zanon, son of physician and engineer Giovanni Antonio Zanon, whose works were on display. In Florence, the Observatory celebrated its 100th anniversary (it had been founded in 1826, actually) through the exhibition of texts published by the institution, historical diagrams and tables on Venice's climate, photographs of its instruments, rain gauges, and a portrait of Massimiliano Tono, director of the Observatory from 1876 to 1902. Given the exiguity of the number of objects presented, the exhibits of Vicenza, Bassano, Belluno, and Treviso were grouped into a single layout which featured scientific texts and guide to collections like Alfredo Jona's *La collezione monumentale di Lazzaro Spallanzani.*

3.2 *Emilia*

The Emilia room hosted the Exhibits of Parma, Modena, Mantua and Lodi, even though the latter two cities are actually situated in Lombardy. Conveniently enough, the room was located between the Milan and Bologna rooms, signaling the gradual shift from Lombardy to Emilia Romagna. The participation of Parma and Modena was coordinated by the Bolognese committee, while the exhibits of Mantua were curated by Giovanni Pinelli, principal of the "Virgilio" High School of Mantua.

Parma presented a collection of 101 objects coming from the Institute of Radiology, the Palatine Library, and the "Maria Luigia" High School. The layout was curated by Pietro Zorzanello, director of the Palatine Library, under the supervision of Antonio Pensa, Rector of the University of Parma.[49] The main theme of the exhibit was history of medicine: there were on display medical manuscripts dating back to the 15th-Century from the Palatine Library and a small collection of surgical instruments belonged to Giovanni Rossi. A special section was devoted to Macedonio Melloni: in a showcase there were on

48 Pietro Pagnini, "Il seminario vescovile di Padova alla I Esposizione Nazionale di storia della scienza," *Unità Cattolica*, June 29, 1929.

49 "La partecipazione di Parma alla Mostra fiorentina," *Corriere Padano*, May 12, 1929.

display several of his instruments and devices. The exhibit of Modena, curated by professor of Physics Carlo Bonacini, consisted of 28 objects. It spanned from Medicine and Botanic to Physics and Astronomy, for Bonacini managed to gather medical manuscripts held by the *Estense* Library; herbals compiled by Casapini, Morelli, and Volpari; a galvanometer and publications on electricity by Marianini; and a telescope built by Giovanni Battista Amici.[50]

Mantua's exhibit, as said, was curated by Giovanni Pinelli. To stage it, Pinelli borrowed Paolo Barbieri's herbals from the Natural History Museum and some electrical machines used by Nobili and Zamboni from the Physics Cabinet of the High School he presided. He also borrowed botanical texts, surgical instruments, and medical consults.[51] As for Lodi, the city presented 9 items: papers on vulcanology and electricity, celebratory posters of botanist Agostino Bassi (who was a native of Mairago, a small town in the nearby of Lodi), and two microscopes built by Giovanni Cavalieri and held by the Barnabite College of Lodi.

3.3 *Campania*

The local committee established in Naples in 1927 was headed by mathematician and historian of science Roberto Marcolongo, and featured, among others, superintendent Amedeo Maiuri and General Inspector of Ancient Harbours and Maritime Works, Luigi Jacono. They selected 81 objects, most of them printed texts. Among these there were a collection of 15 ancient editions of Della Porta's works and a selection of papers by Gaetano Corrado, pioneer of Legal Medicine in Italy. Portraits of Neapolitan scientists were attached to the walls, while the room was occupied by a scale model of a Pompeiian villa and instruments, tools and models of machines from Roman times. Two gromas were on display in the rooms of Rome and of the Military Geographic Institute, respectively, whilst Marcolongo's model of the instrument used by Leonardo to demonstrate Alhazen's theorem was on display in the Tuscany room. After the Exhibition closed, Maiuri and Jacono managed to establish a section of 'ancient technology and mechanics' within the National Museum of Naples. The new section, inaugurated on December 3, 1932, featured the models built for the exhibition of 1929 and new ones executed for the occasion. Among them, there were hand- and horse-mill-stones, parts of the *fistulae* which used

---

50 "La partecipazione di Bologna all'Esposizione di storia della scienza," *il Resto del Carlino*, August 8, 1927.

51 "La partecipazione mantovana alla mostra delle scienze a Firenze," *la voce di Mantova*, May 14, 1929.

to carry water to *Calidarii*, models of olive-presses, *epistomia* (faucets), and a scale model of a Roman pool.[52]

3.4 *Lunigiana*

The city of La Spezia, which had been constituted as Capital of Province in 1923, participated at the Exhibition on behalf of the historic region of Lunigiana. The local committee was headed by Saverio Nasalli Rocca, *podestà* of La Spezia, and the exhibit was curated by Carlo Caselli, curator of the Civic Museum of La Spezia. Eugenio Pedrinelli, director of the Municipal Office for Public Instruction of La Spezia, served as Secretary. The local committee intended the participation at the Exhibition as an occasion to affirm the role of Lunigiana in the Italian scientific tradition and, through it, to reinforce the regional sentiment already acknowledged by the fascist regime in 1923. A brief guide provisionally titled *La Spezia e la Lunigiana nella storia della scienza* ("La Spezia and Lunigiana in the history of science") was prepared and published in 1929 with the title "Bio-bibliographic Essay on the scientists of Lunigiana."[53] The exhibit was housed in the Palace of Expositions' smallest room and consisted of 77 objects, most of them manuscripts and printed works. A collection of 12 portraits of Lunigianese scientists was also on display. The main figures celebrated in the layout were those of botanist Domenico Viviani, chemist Cesare Bertagnini, and naturalist Girolamo Guidoni. Actually, also the Technical Naval Museum of Spezia participated at the Exhibition, but, being an institution directly administered by the Navy, its collections were put on display in the Pavillion of the Ministry of the Navy.[54]

4 Ground Floor IV

4.1 *Ministry of Communications*

The exhibition staged by the Ministry of Communication and curated by Ettore Baldacci, Panfilo La Civita, and Alfredo Rossi spanned four rooms at the farthest end of the Palace of Expositions' ground floor.[55] It largely consisted of relics belonging to Telegraphic Museum of Rome managed by the Ministry

---

52 Amedeo Maiuri, "Una sezione di tecnologia e meccanica antica nel Museo Nazionale di Napoli," *L'Illustrazione Italiana*, n. 46, November 13, 1932; Roberto Minervini, "Una mostra di macchine e strumenti antichi nel Museo Nazionale di Napoli," *il Lavoro fascista*, December 4, 1932.

53 *Saggio bio-bibliografico degli scienziati della Lunigiana* (La Spezia: tip. Moderna, 1929).

54 "La partecipazione di Spezia all'Esposizione di Storia e Scienza," *Telegrafo*, September 2, 1928.

55 "Il padiglione delle Comunicazioni alla Mostra della Scienza," *il Corriere d'Italia*, March 1, 1929.

itself. The Museum was created in 1878 by Ernesto D'Amico and was originally located in Florence. In 1907, it was put under the administration of the Ministry of Communications and transferred to Rome. The exhibition was divided in two sections. The first one focused on history and was loosely centered on the figure of Alessandro Volta, whose sculpture stood in the center of the main room. It was meant to depict the development of telegraphic communications from their origins up to Marconi's wireless systems. In the room occupied by this section there were on display several models of telegraphs and old telephones, telegraphic cables, batteries, and photographs of telegraphic systems. The second section was devoted to "modern systems" and was dedicated to Giovanni Caselli, the inventor of the so-called Pantelegraph, the first device that could telegraphically transmit handwriting and small drawings. Models of the pantelegraph, photographs, sketches, diagrams, and a portrait of Caselli himself were on display. A large part of the exhibit consisted of devices and machineries currently in use in Italian post office, complete with explanations of their functioning and insights into Marconi's wireless systems.

4.2 *Vatican State*

To reach the Ministry of Communications room, one ought to pass a narrow corridor where other small exhibits had been set up. The first exhibit to be encountered after exiting from the Turin room was that of the Vatican State. The Vatican Museum sent celestial, terrestrial, and armillary spheres, and a collection of photographs of the Gallery of Maps located in the Vatican.

4.3 *Ministry of Economy*

The Ministry of Economy staged an exhibit consisting of a collection of 32 machines and devices belonging to the Ichthyogenic Centre of Brescia. The institution, which had been founded in 1892 as a fish farming station, had redefined its mission 1921 and since then had been conducting scientific experiments and developing pollution analysis and control policies.

5 Basement

5.1 *Florence*

As the main exhibition of Florence had been staged in the Palace of Exposition's biggest hall on the ground floor, the exhibits of the remaining Florentine institutes were staged in seven rooms in the basement. The exhibits were curated by a local committee featuring representatives of the institutions involved: the Observatory of Arcetri, the University of Florence, the Laboratory of Chemistry, the *Officine Galileo* Company, the *Alla querce* College, the Geographic Military Institute, the Botanical Institute, and the Georgofili Academy.[56]

The first two rooms were occupied by the Geographic Military Institute. The exhibit was curated by Attilio Mori, professor of Cartography at the Institute and director of *Rivista Geografica Italiana* (Italian Geographic Review). In the first room, the Institute exhibited its instruments; in the second, a collection of cartographies and maps. The instruments were arranged by type: goniometers, helioscopes, theodolites, planimeters, and instruments for calculating distance and height. On the overall, the exhibition staged by the Geographic Military Institute was rather consistent and, contrary to the general layout of the Exhibition, was ordered according to typological criteria.

In the following room there were on display the collections of the Observatory of Arcetri, the "Dante" High School, the *Alla querce* College, and the Astronomical Observatory of Collurania (even though the latter is located in Teramo, Abruzzo). The Observatory of Arcetri sent two globes, terrestrial and celestial. They were originally meant to be put on display in the Florence room on the upper floor, but lack of space forced the curators to move them downstairs. The "Dante" High School sent samples from its precious physics collections dating back to the 19th century,[57] while the *Alla querce* College presented a collection of instruments belonged to barnabites. Among them, an electric lamp built by Giovanni Cavalieri, a coherer by Timoteo Bertelli, and several photographs of the photographic tromometer designed by Camillo Melzi. The Observatory of Collurania sent a portrait of its founder, Vincenzo Cerulli, photographs of the observatory, and an interferometer. The exhibit was completed by an edition of Giovanni Matteo Contarini's geographical map sent by the Wellcome Historical Medical Museum of London. On the opposite side of the above rooms there were, arranged in four more rooms, the exhibits staged by the Georgofili Academy, the Laboratory of Chemistry, the Fabbroni Family's private collection, the University, the Museum of Natural History, and the Ximenian Observatory. The Georgofili Academy presented a collection of manuscripts. The Laboratory of Chemistry, directed by Luigi Rolla, exhibited 42 small jars containing chemical preparations and a collection of instruments and relics variously related to Ugo Schiff, pioneer of the Florentine chemical tradition. The exhibit on Giovanni and Adamo Fabbroni, staged thanks to the

56 "Una esposizione nazionale di storia della scienza a Firenze," *la Nazione*, August 6, 1927.

57 See Maria Teresa Aristodemo, *La collezione dell'antico gabinetto di Fisica del Liceo Dante di Firenze*, in *Convegno Nazionale su strumenti e cultura scientifica nell'Ottocento in Italia (Firenze, 3–4 maggio 1995)* edited by Emilio Borchi, Renzo Macii, and Flavio Vetrano (Firenze: Idealpress, 1997).

contribution of the Fabbroni Family, consisted of portraits, texts, and documents. The exhibit of the University of Florence featured collections and relics coming from the Institute of Pathological Anatomy, the Medical Library, the Institute of Pharmacology, the Museum of Anthropology, the Institute of Anatomy, and the Institute of History of Science. The Institute of Pathological Anatomy celebrated the figure of Guido Banti (who had died in 1925) by showing a series of diapositives on the syndrome bearing his name. The Medical Library exhibited a collection of medical works by Ficino, Tagliacozzi, Fabrici d'Acquapendente; and manuscripts and portraits of Redi, Mascagni, Chiarugi, Betti, and Puccinotti. The exhibit staged by the Institute of Pharmacology focused on the drugs brought to Florence from Southern America by Paolo Mantegazza, whose figure was also celebrated by the exhibit of the Museum of Anthropology. Finally, the exhibit of the Institute of Anatomy consisted of anatomical waxes by Pacini, Mascagni, Paganucci, and Bianchi. The Museum of Natural History, which, at the time of the *Regio Istituto di Studi Superiori*, used to own a huge amount of artifacts and historic relics,[58] staged an exhibit focused on zoology, comparative anatomy, botany, and mineralogy. Zoology was represented by a series of bestiaries and early modern treatises on animals; comparative anatomy by Susini's and Calenzuoli's anatomical waxes of human and cat. The botanical exhibit celebrated Filippo Parlatore by exhibiting his 10-volume treatise *Flora Italiana* (Italian Flora), whilst the mineralogical section featured treatises by Agricola and de Boodt. The Florentine exhibition was completed by the exhibit staged by the Ximenian Observatory of Florence, which sent a selection of instruments and devices used for astronomical and seismological researches, and materials on the Barsanti-Matteucci fuel engine. As the Observatory could not suspend meteorological observations, none of the Observatory's meteorological instruments was sent to the Exhibition.

5.2 *Trento*

The city of Trento had been annexed to the Italian Kingdom only 10 years before the Exhibition and by then had rapidly became the main center of a movement of "Italianization" of Trentino and German-speaking Alto Adige backed and fostered by the fascist regime. The participation of Trento to the Exhibition was championed by Trentinese Senator Vittorio Zippel and managed by the Natural History Museum of Trento. The exhibit was curated by Ezio Mosna,[59] a geography and history enthusiast, leader of the so-called *Legione Trentina* (Trentin Legion), the association of Trentinese soldiers who,

58 See ch. 1.

59 "La nostra Regione all'Esposizione di Storia della Scienza," *Brennero*, May 9, 1929.

during the First World War, had decided to join up with the Italian Army. The *Legione Trentina* was the main force behind the "Italianization" movement, anticipating the systematic measures taken by the fascist regime. In his work, Mosna was adjuvated by naturalist Lino Bonomi, who, after the Exhibition closed, published a guide titled *Naturalisti, medici e tecnici trentini. Contributo alla storia della scienza* ("Trentinese naturalists, physician, and technicians. A contribution to history of science").

Mosna's main aim, however, was to affirm "the active Italianity" of Trento by means of "the most irrefutable documentation of ancestry" which would reveal "all those very Italian names, bound by historical tradition to the name of Italy, and consecrated to the religion of science."[60] The exhibit consisted of 244 objects and, most notably, was the only one that retained an order similar to the Exhibition's original one, for it was structured in five sections: I. Zoology, botany, mineralogy, geology, museums, societies, academies; II. Medical sciences; III. Mathematics, physics, chemistry; IV. Astronomy and Geography; V. Technology. However, ultimately, the room was a gallery of scientists who were born or had lived in Trento and Trentino. Visitors could admire portraits and works of Felice Fontana, Piero Andrea Mattioli, Giovanni Antonio Scopoli, Giacomo Bresadola, and Cesare Battisti (Section I); Giulio Alessandrini, Ippolito Guarinoni, Ottaviano Rovereti, and Giovanni Battista Borsieri (Section II); Gregorio Fontana, Francesco Malfatti, Eustachio Chini, Martino Martini (Sections III and IV); Luigi Negrelli and Giovanni Dall'Armi (Section V). In 1939, the exhibition was restaged at the Natural History Museum of Trento in order to celebrate the 20th anniversary of the annexation and the 10th of the participation at the Exhibition of History of Science.

5.3 *Romagna*

The historic region of Romagna was represented by the cities of Imola, Lugo, Forlì, Ravenna, Faenza, and Ancona (though, strictly speaking, Ancona is not located in Romagna). The exhibits were staged in a small room and as a whole there were about 200 objects on display.

The exhibit of Imola and Lugo, which staged a single exhibition, was curated by Romeo Galli, Head Librarian at Imola's Public Library, and Giuliano Mambelli, director of the Public Library and Historic Archive of the City of Lugo.[61] Imola contributed with 101 items, most of them scientific printed books

60 "Alla prima Esposizione Nazionale di storia delle scienze. Il documento dell'attiva italianità del Trentino," *Brennero*, May 12, 1929; and "Le città redente alla mostra della scienza. Il contributo del Trentino," *Brennero*, October 16, 1929.

61 "Alla mostra della storia delle scienze," *Diario*, October 19, 1929.

of the 18th and 19th century, while Lugo celebrated the figure of Agostino Codazzi, the Lughese geographer, cartographer, and explorer renowned for his explorations of Southern America. The city of Forlì participated at the Exhibition through its Public Library (directed by Benedetto Pergoli) and the Agricultural Chemistry Laboratory. The former sent writings, documents, and portraits commemorating Giovan Battista Morgagni, Maurizio Bufalini, and Carlo Matteucci; the latter presented a series of studies carried out in the Institute since its foundation in 1872. The Public Library of Ravenna sent a couple of works by zoologist Giuseppe Ginanni.[62] In Ancona, a local committee was established in 1927. It was headed by director of "Umberto I" General Hospital, Umberto Baccarani, and featured, among others, Giuseppe Moretti (superintendent of Museums and Antiquities of Marche and Abruzzi), and Luigi Serra (director of the National Galleries of Marche and superintendent of Medieval and Modern Art). The exhibit consisted of four celestial and terrestrial globes, a pendulum, and works by mathematician, engineer and physicist Carlo Rinaldini.[63] The exhibit of Faenza, curated by Pietro Zama (director of the Civic Library of Faenza), was centered around the figure of Evangelista Torricelli. The City of Faenza had already staged an exhibition on Torricelli in 1908, and in 1923 Zama had managed to establish a small museum within the Civic Library.[64] Once again, Zama intended to reassert the claim that Faenza (and not Imola) was Torricelli's homeland. There were on display relics and replicas of Torricelli's instruments whose originals were held in Galileo's Tribune, a portrait of Torricelli by Giovanni Piancastelli, Torricelli's manuscripts, and the three-volume edition of Torricelli's *Opere* edited by Gino Loria and Giuseppe Vassura. A collection of scientific works belonging to the Civic Library of Faenza completed the exhibit.

5.4 *Others*

Under the name "Others" (or also "Various regions") were grouped together a number of exhibits staged by various Italian cities and housed in four small rooms. They consisted, for the most part, of small collections of texts, books, and documents. The cities were: Catania, Siracusa, Cagliari, Conegliano, San Remo, Camerino, Ascoli Piceno, Alanno, Todi, Crotone, Urbino, Macerata, Foggia, Perugia, Gallipoli, Molfetta, Catanzaro, Reggio Calabria, Rosarno,

62 G. Scaioli, "La biblioteca comunale all'esposizione di storia della scienza," *il Popolo di Romagna*, May 18, 1929.

63 "Mostra nazionale di storia della scienza," *il Corriere Adriatico*, December 8, 1927; and "Ancona e l'Esposizione di Firenze," *il Corriere Adriatico*, May 10, 1929.

64 See ch. 1.

Alatri, Fermo, Spoleto, Lecce, Bari, Noicattaro, Lenola, Salerno, Capodistria, Palermo, Idria, Trieste, Postumia, and Zara.

In Catania, a local committee presided by zoologist Achille Russo selected and sent to Florence 63 objects belonging to the University's Library. The exhibit, curated by Botanist Emilio Chiovenda, was aimed at demonstrating the vitality of the University of Catania, which, after the Gentile Reform, saw its subsidies drop. Apart from printed books, there were on display gregorian telescopes, herbals, and surgical instruments. A sundial coming from Siracusa was also among the Catanese artifacts.[65] Catania shared its room with Cagliari, Conegliano, and Sanremo. Cagliari sent just 12 photographs of Nuragic constructions; Conegliano celebrated its School of Enology and Viticulture with documents, photographs, and texts by and on Gera, Carpené, and Caronelli firms; Sanremo sent Cassini's astronomical tables of 1740.

The next room housed diverse collections coming from 21 cities. The exhibit of Camerino, curated by a local committee headed by chemist Pietro Saccardi, was centered on the University; that of Ascoli Piceno, staged by the City Council, commemorated Cecco d'Ascoli, Giovanni Tranquilli, and Antonio Orsini; Alanno presented the activity of the Practical School of Agricolture; thanks to Umberto Bartolini, Todi exhibited drawings and writings by Abdon Menicali, pioneer of aviation. The participation of Macerata, Foggia, Perugia, and Molfetta were promoted by the City Councils of those cities, and mostly consisted of celebrations of individuals: Antonio Pizzarello, Domenico Cotugno, Egnazio Danti, Giuseppe Maria Giovene, etc. The exhibit of Gallipoli, curated by the local section of the Fascist Party presided by Guido Franco, celebrated Giovanni Presta and Antonio De Pace.[66]

The exhibits of Reggio Calabria, Catanzaro, Cosenza, Rosarno, and Crotone were curated by a local committee whose constitution had been prompted by Bellarmino Pasquino in 1927 and which was headed by Francesco Geraci and Vittorio Visalli.[67] While the exhibit of Reggio Calabria celebrated the figures of Placido Geraci and Raffafele Piria, Filippo De Nobili, director of the Civic Library of Catanzaro, curated a small exhibit on seismology. The Nunziante Family of Rosarno sent documents and photographs of the reclamation of the lands in the nearbies of Rosarno; the School of Practical Agricolture of Cosenza and the Cosentine Academy sent books and texts representing their

65 "I Esposizione nazionale di storia delle scienze a Firenze," *il Corriere di Catania*, November 22, 1927.

66 On these exhibits, see: "Prima Esposizione Nazionale di storia della scienza," *In Cammino*, December 27, 1927 (Camerino); "Da Gallipoli," *Gazzetta del Mezzogiorno*, April 28, 1929.

67 "La Calabria all'Esposizione di storia delle scienze," *il Giornale d'Italia*, November 22, 1927; see also "L'Accademia Cosentina all'Esposizione Nazionale di Storia della Scienza," *Cronaca di Calabria*, November 7, 1929.

holdings and activity. From Crotone, Raffaele Lucente sent a manuscript by his father, the physician of the same name. An analogous "regional" committee was established in Puglia to coordinate the participation of Lecce, Bari and Noicattaro. It was presided by Sebastiano Apostolico Orsini Ducas, president of the Association of the Museums of Lecce, who was then busy trying to promote the constitution of a new region – called Salento – separated from Puglia. Lecce presented works by Giorgio Baglivi, Cosimo Moschettini, Giuseppe Palmieri, Gaetano Stella, Giulio Cesare Vanini, and others; Bari sent portraits and works by Giuseppe Saverio Poli and Giuseppe Maria Giovene, and a collection of 62 biographies of Pugliese scientists; Noicattaro exhibited items related to Francesco De Blasio and Giuseppe Logroscino.[68] Alatri, Fermo, Lenola, Capodistria, Idria, and Salerno staged their exhibits in a small room shared with Palermo, whose exhibit, curated by Filippo Angelitti, director of the Astronomical Observatory, focused mainly on medicine and botany. The latter was represented by herbals and books such as Pietro Castelli's *Hortus Messanensis* and Francesco Cupani's *Hortus Catholicus*. A small section was devoted to astronomy, mathematics, and natural sciences, with books like Hodierna's *Archimede redivivo* and Borelli's *Historia et meteorologia incendii Aetnaei anni 1669*.

A more ambitious exhibit was that of Trieste, that, in spirit, was similar to that of Trento. Indeed, the local committee presided by Giulio Morpurgo chose "Italianity" as the main guide for the layout curated by Augusto Riccio di Solbrito. The exhibit celebrated figures such as Domenico Lovisato, Antonio Lorenzutti, Carlo de Marchesetti, and even Giuseppe Tartini, a musician and composer. Besides that, there were a section devoted to naval science where nautical instruments and models of ships were on display. Attached to the Trieste room there were the exhibits of Postumia and Zara, which were also centered on "Italianity."

The exhibit staged by the University of Urbino consisted mostly of photographs of the collections owned by the Physics Cabinet and of texts belonging to the University Library. The Physics Cabinet, after having experienced a huge phase of expansion during the 19th century, had rapidly declined, and by 1929 it was *de facto* a warehouse for instruments no longer in use.[69]

68 "Per la mostra delle scienze a Firenze," *Gazzetta del Mezzogiorno*, April 17, 1929.

69 See ch. 1, and Roberto Mantovani and Flavio Vetrano, *Origini e vicende storiche del Gabinetto di Fisica dell'Università di Urbino*, in *L'Università di Urbino 1506–2006. Vol. II: I saperi fra tradizione e innovazione*, edited by Stefano Pivato (Urbino: Università degli Studi di Urbino, 2010), 445–464.

6 Pavilions

6.1 *Ministry of Aeronautics*

The exhibit staged by the Ministry of Aeronautics was housed in two pavilions and was curated by col. Boccacci under the supervision of col. Marelli, director of the Aeronautics Historical Institute of Rome.[70] Throughout the fascist regime, Aeronautics was assigned an increasingly relevant position. Mussolini was passionate about flying and endorsed the myth of Aeronautics as "the most Fascist" among the Italian Military Corps. A "Commission for Aeronautics" had been created in 1923, presided by Mussolini himself, and the same year it was decreed that Aeronautics should be an autonomous Military Corps.

The feats of pilots like Francesco De Pinedo, Mario De Bernardi, Arturo Ferrarin, Carlo Del Prete, and Italo Balbo were celebrated by fascist propaganda as a powerful representation of the regime's basic ideals: vigor, bravery, boundlessness velocity, dynamism. Italian Aviation, according to the Regime, should emerge and perform extraordinary feats to strike the imagination of both Italians and foreigners.[71] At the Exhibition, the Ministry of Aeronautics celebrated such feats with relics, memorabilia, models of airplanes, helicopters, airships, seaplanes, engines, propellers, altimeters, barometers, compasses, navigational instruments, machine guns, rockets, bombs, and parachutes.

6.2 *Ministry of War*

The pavilion assigned to the Ministry of War consisted of a central hall and two rooms. The objects on display were arranged so as to give an overview as complete as possible of the applications of science in warfare. In the main hall there were portable firearms made in Italy, arranged in four sections (Long Guns, Rifles, Automatic Firearms, Handguns) and disposed chronologically. On the walls, there were photographs coming from the Artillery Museum of Turin. The second room housed the exhibit of the Army Corps of Engineers, staged thanks to the photographs sent by Museum of the Army Corps of Engineers of Rome. The third room was devoted to Battlefield medicine and was divided in two sections: field surgery and "field pharmacy." There were studies on the effects of bullets on the human body and a collection of instruments: portables X-ray machines, surgical instruments, sterilizers, and gas masks. The second section featured recipes for making drugs, a small portable lab for chemical analysis, and medication equipment.

---

70 "La mostra dell'Aviazione," *Regime fascista*, April 23, 1929.

71 See *Le ali del ventennio. L'Aviazione italiana dal 1923 al 1945. Bilanci storiografici e prospettive di giudizio*, edited by Massimo Ferrari (Milano: FrancoAngeli, 2005).

6.3 *Ministry of the Navy*

Having at its disposal a Pavilion with five rooms, the Ministry of the Navy staged an exhibition arranged in five sections: models, guns and machines, Hydrographic Institute of Genoa, Technical Office for Lighthouses of La Spezia, and radiotelegraphic devices. A significant number of the items on display came from the Technical Naval Museum of La Spezia.

The first room housed models of ships; the second one, rockets, torpedoes, and bombs. The Hydrographic Institute of Genoa sent part of its collection of nautical instruments (the other being on display in the Genoa Room); the Technical Office for Lighthouses of La Spezia staged an evocative exhibition of instruments which featured a scale model of the Gulf of Naples complete with buoys, and functioning lighthouses.[72] In the fifth room, finally, there were instruments of communication and devices for transmission.

6.4 *Ministry of Public Works*

The exhibition of the Ministry of Public work was curated by Evaristo Zambelli under the supervision of Minister Giovanni Giuriati. The exhibit focused on five major infrastructure works realized in Italy over the centuries and supposedly resumed and enlarged by the fascist regime: hydrographic systems and studies on the course of the Po river; land reclamation in the Val di Chiana; work on the Arno riverbed carried out under the direction of Lorenzo Materassi; land reclamation around Grosseto; and the work of consolidating the banks of the Adige river.[73]

72 Erasmo Dell'Onore, "La Marina da guerra italiana all'Esposizione di storia della scienza," *Telegrafo*, July 14, 1929.

73 "Il Ministero dei Lavori pubblici alla Mostra della Scienza," *il Corriere d'Italia*, May 17, 1929.

Conclusion

At first sight, the First National Exhibition of History of Science was one of the many initiatives meant to spread fascist values, myths, and narratives. Its organizers alternatively presented it as "a display of Italianity," a huge collection of "national priority claims," and a "demonstration of how Italy had been the world's science teacher." A close look at the history and aims of the Exhibition itself, however, reveals a different truth and allows us to appreciate how complex the relationships between cultural initiatives and nationalistic propaganda can be.

The more than nine thousand items on display, coming from eighty cities and more than two hundred public and private lenders, offered the fullest possible overview of the contribution of Italy to scientific progress. The Exhibition raised awareness of the need to protect the scientific heritage, generated support for the establishment of a Museum of the History of Science towards this end, and spurred Universities, libraries, museums, and other institutions to comb their archives for relevant material, thus providing a valuable initial census of Italy's historical scientific heritage. Even though the Exhibition ultimately failed to achieve its most ambitious goals (special superintendencies for history of science were never established and a census of the Italian scientific books and artefacts was not carried out) it left durable traces in the Italian culture, with today's Museo Galileo – a well-recognized international institution and research center – being the most visible of all.

These results, however, could have not possibly been achieved without the fascist regime's political support, but this support obviously came at a cost. From a historical point of view, the Exhibition is an excellent case study of the complex relationships between culture and dictatorship as it offers precious insights into the fascist regime's attitude towards cultural events and how such attitude affected their aims and scope. The Exhibition was without doubt an event marked with political ideology and defined by externally imposed features. The myth of Italian genius, the claim to Italian scientific priorities, the idea of a national tradition made of realism and experimentalism, the centralization of scientific research became, in the short run, its main themes. The event itself, in the eyes of the fascist regime, was nothing more than a propagandistic resource. Once the Exhibition closed, the Government gave no support either to the National Museum of History of Science or to the project of indexing the Italian scientific collections, neither it implemented a protection policy for scientific artifacts.

The effects of such a relationship are apparent in the Exhibition layout. The organizers knew that Mussolini's interest in history of science could fade away

 | DOI:10.1163/9789004297418_010

as quickly as it had appeared, so they tried to complete their goal as fast as possible. They gave up any attempt at presenting a historically sound image of the development of Italian science and chose to arrange the Exhibition geographically. Participating institutions were granted large freedom in setting up their own exhibits and were allowed to push and celebrate local narratives and regional traditions, causing puzzlement in visitors. They would often find themselves in front of mysterious devices, unknown machines, indecipherable manuscripts, books filled with obscure diagrams and figures, and austere portraits of unheard-of people. This was a common fate for many cultural initiatives undertaken during fascism. On the one side, fascism's authoritarianism, nationalism, centralism, and disregard of bureaucracy offered opportunities for developing wide-spreading national initiatives; on the other, the fact that these were considered by the fascist government merely as cultural and propagandistic assets often turned them into ephemeral celebrations doomed to oblivion. If the museum established after the Exhibition survived the fall of fascism, it was because the Exhibition itself had an identity strong enough to resist fascism's typical tendency to turn mutualistic interactions with cultural enterprises into parasitic ones, and rich enough to provide the museum with a set of values and narratives independent from those imbued by fascism.

The mutualistic/parasitic approach used to understand the Exhibition, I think, could be extended to look at other cultural institutions born during fascism, helping us understand the shaping of contemporary Italian cultural panorama as a whole. Depending on the nature of their interaction with 19th-century dominant narratives and values, institutions could either disappear or reshape their role and identity once those same values and narratives went extinct. An eloquent example is the case of the CNR (National Research Council), the leading Italian center for scientific research. When considered in a long-term perspective that traces back its roots and values to Vito Volterra and the foundation of the SIPS,[1] the CNR appears not as a fascist institution that was later 'defascistized', but as the surviving host of a mutualistic interaction that, at some point during the 1930s, was evolving into parasitism. Rethinking cultural institutions' history and identity in terms of survivorship[2] does not mean erasing their historical role in spreading dominant narratives. On the contrary, it means disentangling the factors at play in their formation in order to better understand their role, scope, and activity in cultural production and

---

1 Giovanni Paoloni, Raffaella Simili, *Per una storia del Consiglio Nazionale delle Ricerche* (Roma-Bari: Laterza, 2001).

2 I intend "survivorship" as in Mahmood Mamdani, *Neither Settler nor Native: the Making and Unmaking of Permanent Minorities* (Cambridge: Harvard University Press, 2020).

heritage preservation. This is of critical importance today, as museums, exhibitions, and cultural institutions in general are urged to redefine themselves and reconsider their legacy in response to societal change.

On a different note, the Exhibition and the museum greatly contributed to the understanding of scientific objects as historical evidence and sources of otherwise unavailable information, thus paving the way for the more recent 'material turn' in history of science. When, in 1976, Maria Luisa Righini Bonelli launched the *Annali dell'Istituto e Museo di Storia della Scienza*, it appeared immediately that the journal's twofold commitment to research and preservation was something unprecedented in history of science. Around the *Annali* an international community of scholars variously interested in scientific objects soon gathered together: Silvio A. Bedini, Thomas B. Settle, Marshall Clagett, Stillman Drake, Warren Van Egmond, Paolo Galluzzi, Louis Janin, Marie Boas Hall, Willem D. Hackmann, just to name a few – scholars whose works would prove fundamental in the birth of contemporary material and visual history of science.

Bibliography

Primary Sources

The main source for the study of the First National Exhibition of History of Science is the "ESP29" archive owned by the Museo Galileo of Florence. It consists of both published and unpublished material divided in 10 series:

a. *Newsletters*;
b. *Paperworks*;
c. *Organisation and Setup*;
d. *Arrivals, lenders, and participating cities records*;
e. *Ceremonies and other events*;
f. *Press review*;
g. *Preparatory materials for the Great Catalogue*;
h. *Preparatory materials for the Catalogue of 1952*;
i. *Alinari's photographic records – instruments and machines*;
j. *Alinari's photographic records – portraits of scientists.*

The inventory of the ESP29 archive can be seen on the Museo Galileo's website (http://biblioteca.imss.fi.it/856/archivio_esp29_documentazione_testuale.pdf).

A second source is the "Piero Ginori Conti archive" held in the Archivio di Stato di Firenze (Florence State Archive). It consists of 22 boxes whose content is still waiting to be fully indexed and inventoried and which features Piero Ginori Conti's personal documents and letters.

The Archivio Storico del Comune di Firenze (Historical Archive of the City of Florence) owns documents related to the organisation and financial management of the Exhibition. These documents, gathered by Raffaello Bacci, have been digitised by Museo Galileo and are now available on the Museo Galileo website (http://bibdig.museogalileo.it/Teca/Viewer?an=000000983646).

Other materials are held in the University of Florence Archive, the Alinari Archives, and the State Archive of Rome.

Bibliography

"Federazione Nazionale fra le società, gli enti, gli insegnanti ed i cultori di storia della scienza. Comunicazioni ufficiali." *Archivio di storia della scienza* (1924–), *Archeion* (1927–).

"Gruppo per la tutela del patrimonio scientifico nazionale. Comunicazioni ufficiali." *Archivio di storia della scienza* (1923–).

"Istituto Nazionale di Storia della Scienza. Comunicazioni ufficiali." *Archivio di storia della scienza* (1926), *Archeion* (1927–).

 | DOI:10.1163/<DOI>_001

"Memoriale." *Archeion*, 8 (1927): 232–235.

Abetti, Giorgio, "L'edizione nazionale delle Opere di Galileo e la Mostra delle scienze di Firenze," *l'Universo*, 10, 10 (1929), 1007–1014.

Aimone, Linda and Olmo, Carlo, *Le esposizioni universali, 1851–1900: il progresso in scena*. Torino: U, Allemandi, 1990.

Annuario della Stampa Italiana. Vols. 15–16. (Firenze: Casa Editrice del Libro Italiano, 1937).

Aristodemo, Maria Teresa, "La collezione dell'antico gabinetto di Fisica del Liceo Dante di Firenze," in *Convegno Nazionale su strumenti e cultura scientifica nell'Ottocento in Italia (Firenze, 3–4 maggio 1995)*, edited by E. Borchi, R. Macii, F. Vetrano. Firenze: Idealpress, 1997.

Aronsson, Peter; Elgenius, Gabriella (eds.), *National Museums and Nation-Building in Europe, 1750–2010*, New York: Routledge, 2014.

Baculo, Adriana; Gallo, Stefano; and Mangone, Mario (eds.), *Le grandi esposizioni nel mondo, 1851–1900: dall'edificio città alla città di edifici, dal Crystal Palace alla White City*. Napoli: Liguori, 1988.

Barbagli, Fausto, *Il Gioiello, l'ultima dimora di Galileo*. Firenze: Museo di Storia Naturale, 2009.

Baroncelli, Giovanna, and Bucciantini, Massimo, "Per una storia delle istituzioni storico-scientifiche in Italia: l'Istituto e Museo di storia della scienza di Firenze." *Nuncius*, 5, II (1990): 5–52.

Barsanti, Giulio and Chelazzi, Guido, *Il Museo di Storia Naturale dell'Università degli Studi di Firenze*. Volume I. Le collezioni della Specola: Zoologia e cere anatomiche. Firenze: Firenze University Press, 2009.

Barsanti, Giulio; Becagli, Vieri; and Pasta, Renato (eds.), *La politica della scienza: Toscana e stati italiani nel tardo settecento. Atti del convegno di Firenze, 27–29 gennaio 1994*. Firenze: Leo S. Olschki Editore, 1996.

Bassignana, Pier Luigi, *Images du progres: la technique à travers les expositions dans les documents des archives historiques AMMA*. Torino: U. Allemandi, 1990.

Beauchamp, Ken G., *Exhibiting Electricity*. London: The Institution of Elecetrical Engineers, 1997.

Bedini, Silvio A., "The evolution of science museums," *Technology and Culture*, VI, 1 (1965): 1–29.

Bellodi, Giuliano (ed.), *Gli strumenti di Alessandro Volta: il Gabinetto di fisica dell'Università di Pavia*. Milano: Hoepli, 2002.

Bencivenni, Mario and De Vico Fallani, Massimo, *Giardini pubblici a Firenze dall'Ottocento a oggi*. Firenze: EDIFIR, 1998.

Ben-Ghiat, Ruth, *Fascist Modernities: Italy, 1922–1945*, Berkeley-Los Angeles-London: University of California Press, 2001.

Bennet, Tony, *The Birth of the Museum: History, Theory, Politics*. London: Routledge, 1995.

Beretta, Marco (ed.), *From private to public. Natural collections and Museums*. Sagamore Beach: Science History Publications, 2005.

Beretta, Marco, "Andrea Corsini and the Creation of the Museum of the History of Science in Florence (1930–1961)." In *Scientific Instruments on Display*, edited by Silke Ackermann, Richard L. Keremer, and Mara Miniati. Leiden Boston: Brill, 2014: 1–36.

Beretta, Marco, "L'Alma Mater all'Esposizione Nazionale di Storia della Scienza di Firenze (1929)," in *Una scienza bolognese? Figure e percorsi della storiografia della scienza a Bologna dal Risorgimento a oggi*, edited by A. Angelini, M. Beretta, and G. Olmi. Bologna: Bononia University Press, 2015, 263–277.

Beretta, Marco; Pogliano, Claudio; and Redondi, Pietro (eds.), *Journals and History of Science*. Firenze: Leo S. Olschki, 1998.

Bigatti, Giorgio and Onger, Sergio (eds.), *Arti, tecnologia, progetto. Le esposizioni e l'industria in Italia prima dell'Unità*. Milano: FrancoAngeli, 2007.

Boffito, Antonio, *Strumenti della scienza e scienza degli strumenti. Con l'illustrazione della tribuna di Galileo*. Firenze: Barbéra, 1929.

Bònoli, Fabrizio, "Il Museo della Specola del Dipartimento di Astronomia e del Sistema Museale d'Ateneo dell'*Alma Mater Studiorum*," *Annali di Storia delle Università Italiane*, 17 (2007), 489–497.

Bordoni, Carlo, *Cultura e propaganda nell'Italia fascista*. Messina-Firenze: D'Anna, 1974.

Bortolotti, Lando and De Luca, Giuseppe, *Come nasce un'area metropolitana. Firenze Prato Pistoia: 1848–2000*. Firenze: Alinea, 2000.

Bovi, Tina, *Il Museo Storico Nazionale dell'Arte Sanitaria*. Roma: Fratelli Palombi Editori, 2001.

Brenni, Paolo, "Dal Crystal Palace al Palais de l'Optique: la scienza alle esposizioni universali, 1815–1900," *Memoria e Ricerca*, 17 (2004): 35–64.

Bucci, Mario and Bencini, Raffaello, *Palazzi di Firenze*. Firenze: Vallecchi, 1971.

Bucciantini, Massimo, "Galileo e le passioni del Risorgimento," in *Il caso Galileo: una rilettura storica, filosofica, teologica. Convegno internazionale di studi. Firenze, 26–30 maggio 2009*, edited by. Massimo Bucciantini, Michele Camerota, and Franco Giudice. Firenze: Leo S. Olschki 2011: 347–366.

Buscioni, Maria Cristina, *Esposizioni e stile nazionale, 1861–1925: il linguaggio dell'architettura nei padiglioni delle grandi kermess nazionali e internazionali*. Firenze: Alinea, 1990.

Cagnetta, Mariella, *Antichisti e Impero fascista*. Bari: Dedalo, 1979.

Calisi, Marinella, "L'origine delle collezioni scientifiche del Museo astronomico e copernicano in Roma alla fine del XIX secolo," *Nuncius*, IV, 2 (1989), 249–302.

Calisi, Marinella, *Storia e strumenti del Museo astronomico e copernicano di Roma: guida alle collezioni*. Roma: Osservatorio astronomico di Roma, 2000.

Campisano, Carla, "La Fiera Internazionale del Libro di Firenze. 1922–1932," *Bollettino di storia dell'editoria in Italia*, XIII, 1 (2007), 19–22.

Campochiaro, Emilia; Pasquini, Lucia; and Millozzi, Alessandra (eds.), *Giovanni Gentile e il Senato. Carteggio (1895–1944)*. Soveria Mannelli: Rubbettino, 2008.

Canali, Stefano, "La medicina," in *Per una storia del Consiglio Nazionale delle Ricerche*, vol. I. Edited by G. Paoloni and R. Simili. Roma-Bari: Laterza, 2001.

Cannistraro, Philip V., *Fascismo e cultura*. Venezia: Marsilio, 1975.

Cannistraro, Philip V., *La fabbrica del consenso. Fascismo e mass-media*. Roma-Bari: Laterza, 1975.

Cantagalli, Roberto, *Storia del fascismo fiorentino. 1919/25*. Firenze: Vallecchi, 1972.

Capanna, Ernesto; Malerba, Giancarla; and Vomero, Vincenzo (eds.), *Musei scientifici universitari: una grande risorsa culturale da valorizzare*, monographic number of *Museologia scientifica*, 7 (2011).

Casadio, Claudio and Gentilini, Anna Rosa (eds.), *L'Esposizione di Faenza del 1908*. Faenza: Casanova, 2008.

Cavalcaselle, Giovanni Battista, *Sulla conservazione dei monumenti ed oggetti di belle arti e sulla riforma dell'insegnamento accademico*. Torino: Tipografia La Subalpina, 1863.

Ciliberto, Michele, "Scienza, filosofia e politica: Federigo Enriques e il neoidealismo italiano." *Studi Storici*, 22 (1981): 861–886.

Cini, Marco "L'ambigua costruzione di un'identità nazionale: le celebrazioni di Galileo Galilei a Pisa (1864)," in *Galileo e Bruno nell'immaginario dei movimenti popolari tra Otto e Novecento*, edited by Franco Bertolucci. Pisa: BFS, 2001: 67–84.

Como e l'Esposizione Voltiana, 1899. Como: Tipografia cooperativa comense, 1899.

Contardi, Simone, *La casa di Salomone a Firenze. L'imperiale e reale Museo di fisica e storia naturale*. Firenze: Leo S. Olschki, 2002.

Corsini, Andrea, "La Specola di Firenze," in *la Lettura*, XIV, 5 (1924): 377–382.

Corsini, Andrea, "Per il patrimonio storico-scientifico italiano." *Archivio di storia della scienza*, 5 (1924): 348–355.

Corsini, Andrea, "Relazione sull'importanza e i metodi di divulgare la conoscenza della storia delle scienze in genere e della medicina in ispecie," in *Atti del I Congresso nazionale della Società italiana di storia critica delle scienze mediche e naturali, Roma, 1912*. Extract. Grottaferrata: Tip. italo-orientale S. Nilo, 1913.

Corsini, Andrea, "Firenze e la storia delle scienze." *Atti della società colombaria di Firenze*, February 20, 1927: 256–263.

Corsini, Andrea, "L'Istituto di Storia della Medicina nella Università di Lipsia." *Rivista di storia delle scienze mediche e naturali* (1919): 165–166.

Cresti, Carlo, *Firenze, capitale mancata. Architettura e città dal piano Poggi a oggi.* Milano: Mondadori Electa, 1997.

De Benedictis, Cristina; Giusti, Giovanna; Miniati, Mara; and Pallotta, Eva (eds.), *Presenze femminili nella cultura del Novecento: Adriana Tramontano, Maria Luisa Righini Bonelli, Luisa Becherucci.* Firenze: Centro Di, 2008.

De Felice, Renzo, *Mussolini il duce. I. Gli anni del consenso. 1929–1936.* Torino: Einaudi, 1974.

De Felice, Renzo, *Mussolini il fascista. I. La conquista del potere.* Torino: Einaudi, 1966.

De Felice, Renzo, *Mussolini il fascista. II. L'organizzazione dello Stato fascista.* Torino: Einaudi, 1969.

De Grazia, Victoria, *The Culture of Consent: Mass Organization of Leisure in Fascist Italy.* New York: Cambridge University Press, 2010.

De Gregori, Giorgio and Buttò, Simonetta, *Per una storia dei bibliotecari italiani del XX secolo. Dizionario bio-bibliografico 1900–1990.* Roma: AIB Associazione Italiana Biblioteche, 1999.

Decleva, Enrico; Lacaita, Carlo G.; and Ventura, Angelo (eds.), *Innovazione e modernizzazione in Italia fra Otto e Novecento.* Milano: FrancoAngeli, 1995.

Del Lungo, Carlo, "Ancora a proposito del Museo di Fisica e Storia Naturale di Firenze." *Il Marzocco*, 27, n. 35, August 27, 1922.

Del Lungo, Carlo, "Il R. Museo di Fisica e storia naturale di Firenze: quello che è e quello che dovrebbe essere." *il Marzocco*, 27, n. 33, August 13, 1922.

Dessì, Paola, "Le origini della storia della scienza: discussioni italiane del primo Novecento," *Giornale critico della filosofia italiana*, LXXXII/LXXXIV, 23, n. 2 (2003), 254–272.

Devincenzi, Giuseppe, *Del Museo Industriale Italiano e del progetto di legge pel suo ordinamento. Osservazioni di G. De Vincenzi.* Torino: Tipografia di Enrico Dalmazzo, 1865.

Di Giovanni, Marco, *Scienza e potenza: i miti della guerra moderna, istituzioni scientifiche e politiche di massa nell'Italia fascista. 1935–1945.* Torino: Zamorani, 2005.

Dizionario Biografico degli Italiani. Accessed July, 2015. http://www.treccani.it/biografie/.

Edwards, Anthony D., *The Role of International Exhibitions in Britain. 1850–1910. Perceptions of Economic Decline and the Technical Education Issue.* Amherst: Cambria Press, 2008.

Emiliani, Andrea, "Musei e Museologia," in *Storia d'Italia, i documenti, V*, 1615–1655. Torino: Einaudi, 1973.

Enriques, Federigo, *Il significato della storia del pensiero scientifico.* Manduria: Barbieri, 2005.

Ernesti, Giulio, "La grande città nel ventennio fascista," in *La grande Genova 1926–2006. Atti del Convegno di Studi, Genova, 28–30 novembre 2006*, edited by E. Ariotti, L. Canepa, R. Ponte. Genova, 2008, 117–144.

Fabbro, Mario, *Fascismo e lotta politica in Friuli (1920–1926)*, Venezia: Marsilio, 1974.

Fantozzi Micali, Osanna, "La tutela del patrimonio culturale fra istituzioni e cittadini." In *Beni culturali, ambiente, paesaggio. L'operato delle associazioni nell'azione di tutela*, edited by Osanna Fantozzi Micali and Elena Lolli, 5–10. Firenze: Alinea Editrice, 2009.

Fantozzi, Federigo, *Nuova guida ovvero Descrizione storico-artistica critica della città e contorni di Firenze compilata da Federigo Fantozzi architetto*. Firenze: presso Giuseppe Ducci, 1847.

Favaro, Antonio, "La casa di Arcetri dove spirò Galileo «Monumento Nazionale»." *Archivio di storia della scienza*, 2 (1920), 438–440.

Ferraresi, "Il Gabinetto di Fisica sperimentale nella seconda metà del secolo XVIII: didattica, divulgazione, ricerca nella politica asburgica della scienza," in *Annali delle Università Italiane*, 7 (2003): 91–110.

Ferrari, Massimo (ed.), *Le ali del ventennio. L'aviazione italiana dal 1923 al 1945. Bilanci storiografici e prospettive di giudizio*. Milano: FrancoAngeli, 2005.

Fogu, Claudio, *The Historic Imaginary: Politics of History in Fascist Italy*. Toronto: Toronto University Press, 2003.

Fregonese, Lucio, "Il Museo per la storia dell'Università di Pavia: storia, patrimonio e nuovi allestimenti," *Annali di storia delle Università Italiane*, 5 (2001).

Galluzzi, Paolo, "Gli istituti e i musei di storia della scienza," *Accademie e biblioteche d'Italia*, 52, 4–5 (1984), 416–421.

Galluzzi, Paolo, "L'Accademia del Cimento: gusti del principe, filosofia e ideologia dell'esperimento," *Quaderni storici*, 16, n. 48, 3 (1981).

Galluzzi, Paolo, "La storia della scienza nell'E42," in *E42. Utopia e scenario del regime. Ideologia e programma dell'Olimpiade della civiltà*. Edited by T. Gregory. Venezia: Marsilio, 1987, 53–58.

Galluzzi, Paolo, "La storia della scienza per la diffusione della cultura scientifica," in *La diffusione della cultura scientifica. Giornata Lincea indetta in occasione della III Settimana della Cultura Scientifica e Tecnologica (Roma, 23 aprile 1993)*. Roma: Accademia Nazionale dei Lincei, 1994: 25–35.

Galluzzi, Paolo, "Una rivalutazione idealistica della scienza," *Intersezioni*, 1, 3 (1981), 651–657.

Gamba, Aldo, and Schiera, Pierangelo, *Fascismo e Scienza. Le celebrazioni voltiane e il Congresso Internazionale dei Fisici del 1927*. Bologna: il Mulino, 2005.

Gandolfo, Francesca, *Il Museo Coloniale di Roma (1904–1971). Fra le zebre nel paese dell'olio di ricino* (Roma: Gangemi editore, 2005).

Ganz, Cheryl R., *The 1933 Chicago World's Fair: A Century of Progress*. Chicago: University of Illinois Press, 2012.

Garbasso, Antonio, "L'orma di Galileo: una visita ai laboratori di Arcetri." *Nuovo Patto*, 5 (1922).

Garbasso, Antonio, "La scienza e la filosofia nostra nazionale. Discorso letto all'adunanza solenne del 4 giugno 1922 dal socio Antonio Garbasso." In *Atti dell'Accademia Nazionale dei Lincei*, 3, VIII [excerpt].

Garbasso, Antonio, "La tradizione del pensiero toscano: discorso inaugurale letto nella sala di Luca Giordano nel palazzo mediceo Riccardi il 6 novembre 1916 dal prof. Antonio Garbasso." In *Annuario del R. Istituto di Studi Superiori pratici e di perfezionamento* [excerpt].

Garbasso, Antonio, "Parole del prof. Garbasso." In *Atti della società italiana per il progresso delle scienze, VI riunione* [excerpt].

Garbasso, Antonio, "Per il R. Museo di Fisica e Storia Naturale di Firenze." *Il Marzocco*, 27, n. 35, August 20, 1922.

Garin, Eugenio, *Cronache di filosofia italiana*. Bari: Laterza, 1955.

Garin, Eugenio, *Filosofia e scienze nel Novecento*. Roma: Laterza, 1978.

Gentile, Emilio, *Fascismo. Storia e interpretazione*, Roma: Laterza, 2005.

Gentile, Emilio, *Le origini dell'ideologia fascista*, Bologna: il Mulino, 2011.

Geppert, Alexander C.T. and Baioni, Massimo (eds.), *Esposizioni in Europa fra Otto e Novecento. Spazi, organizzazione, rappresentazioni*. Milano: FrancoAngeli, 2004.

Ghinassi, Giovanni *Per l'innalzamento della statua di Evangelista Torricelli in Faenza nel CCLVI della sua nascita*. Faenza: dalla Tipografia di Pietro Conti, 1864.

Giacomelli, Raffaele, "I modelli delle macchine volanti di Leonardo da Vinci,". *l'Ingegnere*, 2, V (1931).

Giacosa, Pietro, *Magistri salernitani nondum editi: catalogo ragionato dell'Esposizione di storia della medicina aperta a Torino nel 1898*. Torino: f.lli Bocca, 1901.

Giacosa, Pietro, *Notizie sugli oggetti esposti alla mostra di storia della medicina*. Torino: f.lli Bocca, 1898.

Giorgi, Pamela (ed.), *Radici di futuro: l'innovazione a scuola attraverso i 90 anni dell'Indire. Catalogo della mostra* (Firenze: Indire, 2015).

Govoni, Paola, *Un pubblico per la scienza. La divulgazione scientifica nell'Italia in formazione*. Roma: Carocci, 2002.

Gravina, Francesco (ed.), *Le meraviglie dell'ingegno: strumenti scientifici dai Medici ai Lorena*. Firenze: Ponte alle Grazie, 1990.

Greenberg, Reesa; Ferguson, Bruce; and Nairne, Sandy (eds.), *Thinking about Exhibitions*. Routledge: London, 1996.

Hamilton, Alastair, *The Appeal of Fascism: A Study of Intellectuals and Fascism, 1919–1945*, London: Blond, 1971.

Hobsbawm, Eric and Ranger, Terence, *The Invention of Tradition*, Cambridge: Cambridge University Press, 1983.

Israel, Giorgio, "Vito Volterra e la riforma scolastica Gentile," *Bollettino dell'Unione Matematica Italiana*, serie 8, n. 3 (1998), pp. 269–287.

Kranzberg, Melvin and Pursell, Carroll W. (eds.), *Technology in Western Civilization.* New York: Oxford University Press, 1967.

L'Italia alla Esposizione Universale di Parigi del 1867. Rassegna critica descrittiva illustrata. Paris-Firenze: n.p., 1867.

Levi, Donata, "Il viaggio di Morelli e di Cavalcaselle nelle Marche e nell'Umbria," in *Giovanni Morelli e la cultura dei conoscitori*, edited by Giacomo Agosti, Maria Elisabetta Manca, and Matteo Panzeri, 133–148. Bergamo: Lubrina, 1993.

Levi, Donata, *Cavalcaselle. Il pionere della conservazione dell'Arte.* Torino: Einaudi, 1988.

Libri, Guglielmo, *Histoire des sciences mathématiques en Italie.* Paris: Chez Jules Renouard et Cie, 1838.

Linguerri, Sandra, "La Società Italiana per il Progresso delle Scienze: 1909–1930," *Nuncius*, XV (2000): 51–78.

Lohr, Lenox R., *Fair Management. The Story of* A Century of Progress Exposition. Chicago: Cuneo Press, 1952.

Lucci, Leonardo; Salvini, Antonio; Temporelli, Massimo; Valotti, Barbara, "Musei, collezioni e fonti documentali per la storia delle telecomunicazioni in Italia," in *Storia delle telecomunicazioni*, ed. by Gabriele Falciasecca, Virginio Cantoni, Giuseppe Pelosi (Firenze: Firenze University Press, 2011), pp. 939–963.

Lumley, Robert (ed.), *The Museum Time-Machine: Putting Cultures on Display.* London: Routledge, 1988.

Lungonelli, Michele, and Migliorini, Martina, *Piero Ginori Conti. Scienza, cultura e innovazione nella Toscana del Novecento.* Roma-Bari: Laterza, 2003.

Luperini, Claudio and Paladini, Tiziana, *Ho costruito il seguente apparecchietto. Antonio Pacinotti, manoscritti e strumenti.* Pisa: ETS, 2007.

Lussu, Emilio, *Marcia su Roma e dintorni* (Milano: Mondadori, 1968).

Lyttelton, Adrian, *The Seizure of Power: Fascism in Italy, 1919–1929*, London: Weidenfeld & Nicholson, 1973.

Maccagni, Carlo and Freguglia, Paolo (eds.), *La storia delle scienze.* Busto Arsizio: Bramante, 1989.

MacDonald, Sharon (ed.), *The Politics of Display: Museums, Science, Culture*, London-New York: Routledge, 1998.

MacDonald, Sharon, *Memorylands: Heritage and Identity in Europe Today*, New York: Routledge, 2003.

MacDonald, Sharon, "Museums, national, postnational and transcultural identities," *Museum and Society*, 1 (1), 2003: 1–16.

Maiocchi, Roberto, "Il ruolo delle scienze nello sviluppo industriale italiano," in *Storia d'Italia. Annali 3. Scienza e tecnica.* Edited by G. Micheli. Torino: Einaudi, 1980, 864–999.

Maiocchi, Roberto, *Gli scienziati del Duce: il ruolo dei ricercatori e del CNR nella politica autarchica del fascismo.* Roma: Carocci, 2003.

Maiocchi, Roberto, *Scienza e fascismo*. Bologna: il Mulino, 2004.

Maiocchi, Roberto, "Fascism and Italian science policy," *Science and Power: the historical foundations of research policies in Europe*. Luxembourg: Office for official publications of the European communities, 2000, 179–185.

Majuri, Amedeo, "Una sezione di tecnologia e meccanica antica nel Museo Nazionale di Napoli," *L'Illustrazione Italiana*, 46, November 1932.

Mamdani, Mahmood, *Neither Settler nor Native: the Making and Unmaking of Permanent Minorities*, Cambridge: Harvard University Press, 2020.

Mantovani, Roberto and Vetrano, Flavio, "Origini e vicende storiche del Gabinetto di Fisica dell'Università di Urbino," in *L'Università di Urbino 1506–2006. Vol. II: I saperi fra tradizione e innovazione*, edited by S. Pivato. Urbino: Università degli Studi di Urbino, 2010, 445–464.

Marino, Antonio, *Como al tempo del decoro e dell'orgoglio. L'Esposizione Voltiana, l'incendio e la ricostruzione. Gli occhi del mondo sulla sfida di una piccola città*. Como: Alessandro Dominioni Editore, 2012.

Martínez Garrido, Elisa, "Il primo discorso fascista di Mussolini: la traccia dannunziana," *Cuadernos de Filología Italiana*, 5, 213–229.

McDonald, Sharon (ed.), *The Politics of Display. Museums, Science, Culture*. London: Routledge, 1997.

Metron, "Un Museo Industriale Italiano," *Corriere della sera*, September 5, 1931.

Mieli, Aldo, "Per una biblioteca di storia della scienza: lettera aperta a Giovanni Gentile," *Archivio di storia della scienza*, 4 (1923): 305–306.

Mieli, Aldo, "Fronte Unico." *Archivio di storia della scienza*, 7 (1926): 183–184.

Mieli, Aldo, "L'Istituto di storia delle scienze nell'Università di Firenze." *Archivio di storia della scienza*, 6 (1925): 198–199.

Mieli, Aldo, "Per un'intesa fra gli studiosi e gli enti di storia della scienza." *Archivio di storia della scienza*, 5 (1924): 79–81.

Mieli, Aldo, "Per un'intesa fra le società di storia della scienza e l'Archivio." *Archivio di storia della scienza*, 4 (1923): 109–110.

Minervini, Roberto, "Una mostra di macchine e strumenti antichi al Museo Nazionale di Napoli," *il Lavoro fascista*, December 4, 1932.

Miniati, Mara and Contardi, Simone, "Il Museo di fisica e storia naturale," in *La corte in archivio: apparati, cultura, arte e spettacoli alla corte lorenese di Toscana*. Livorno: Sillabe, 1997, 49–58.

Miniati, Mara, "Catalogo del fondo antico dei volumi dell'Istituto di fisica dell'Università di Firenze," *Annali dell'Istituto e Museo di Storia della Scienza di Firenze*, VI, 1 (1981), 225–240.

Miniati, Mara, "Formazione della collezione dell'Istituto e Museo di storia della scienza di Firenze," in *Strumenti di fisica e cultura scientifica nell'Ottocento in Italia*. Firenze: Litografia I.P., 1993, 13–16.

Miniati, Mara, "La collezione di chimica del Museo di storia della scienza di Firenze," in *Atti del III convegno nazionale di storia e fondamenti della chimica*, edited by F. Abbri and F. Crispini. Cosenza: Brenner, 1991, 93–99.

Miniati, Mara, "Origini della Specola fiorentina," *Giornale di astronomia*, x, 3–4 (1983), 209–220.

Miniati, Mara, "Per una storia dell'Istituto e Museo di storia della scienza di Firenze," *Giornale di astronomia*, xxv, 4 (1999), 53–56.

Miniati, Mara, *Museo di storia della scienza: catalogo*. Firenze: Giunti (1991).

Moggi, Giorgio and Roggi, Piero (eds.), *Firenze 1815–1945: un bilancio storiografico*. Firenze: Le Monnier, 1990.

Monciatti, Alessio, *Alle origini dell'Arte nostra. La Mostra Giottesca del 1937 a Firenze*. Milano: il Saggiatore, 2010.

Monza, Francesca and Barbagli, Fausto, *La scienza nei musei: guida alla scoperta dello straordinario patrimonio museale scientifico italiano: musei naturalistici, musei tecnico scientifici, orti botanici, science center, musei di medicina, giardini zoologici, acquari*. Milano: Orme, 2006.

Morghen, Raffaello, *L'Accademia Nazionale dei Lincei nel CCCLXVIII anno dalla sua fondazione nella vita e nella cultura dell'Italia unita (1871–1971)*. Roma: Accademia Nazionale dei Lincei, 1972.

Ojetti, Ugo, *Cose viste. 1928–1930, tomo quinto*. Milano: A. Mondadori, 1937.

Onger, Sergio, *Verso la modernità: i bresciani e le esposizioni industriali, 1800–1915*. Milano: FrancoAngeli, 2010.

Palla, Marco, *Firenze nel regime fascista. 1929–1934*. Firenze: Leo S. Olschki, 1978.

Palla, Marco, *Mussolini e il fascismo*. Firenze: Giunti, 1993.

Paoloni, Giovanni and Simili, Raffaella, *Per una storia del Consiglio Nazionale delle Ricerche*. Roma-Bari: Laterza, 2001.

Papa, Emilio Raffaele, *Fascismo e cultura*. Venezia: Marsilio, 1975.

Partito Nazionale Fascista. Mostra della Rivoluzione Fascista. Inventario. Edited by Gigliola Fioravanti. Roma: Ministero per i Beni Culturali e Ambientali, 1990.

Pavoni, Rosanna and Selvafolta, Ornella, *Milano 1894: le Esposizioni riunite*. Milano: Camera di Commercio, 1994.

Picard, Alfred, *Le bilan d'un siècle*, I. Paris: Imperie Nationale, 1906.

Piccini, Emanuela, "Opere e progetti di Enrico Dante Fantappié (1869–1951). Indagini e osservazioni per la conoscenza e il restauro," *bollettino ingegneri*, 7 (2007), 18–26.

Pigatto, Luisa (ed.), *Giuseppe Toaldoe il suo tempo nel bicentenario della morte: scienza e lumi tra Veneto e Europa*. Cittadella: Bertoncello ArtiGrafiche, 2000.

Piscopio, Ugo, *La scuola del regime: i libri di testo nelle scuole secondarie sotto il fascismo*. Napoli: Alfredo Guida Editore, 2006.

Pogliano, Claudio (ed.), *Scienze e storia nell'Italia del novecento*. Pisa: Plus, 2007.

Pogliano, Claudio and Cassata, Francesco (eds.), *Storia d'Italia. Annali 26. Scienze e cultura dell'Italia unita*. Torino: Einaudi, 2011.

Point, Jean-Claude (ed.), *Pour comprendre le XIXe: histoire et philosophie des sciences à la fin du siècle*. Firenze: Leo S. Olschki, 2007.

Polvani, Giuseppe, *Antonio Pacinotti: la vita e l'opera*. Pisa: Lischi, 1934.

Polvani, Giuseppe, *Le cose storicamente più notevoli possedute dall'Istituto di fisica tecnica della R. scuola d'Ingegneria di Pisa*. Pisa: Pubblicazioni della R. Scuola d'Ingegneria, 1930.

Preziosi, Anna Maria, *Borghesia e fascismo in Friuli negli anni 1920–1922*, Roma: Bonacci, 1980.

Puccini, Sandra, *L'itala gente dalle molte vite. Lamberto Loria e la Mostra di Etnografia italiana del 1911*, Roma: Meltemi editore, 2005.

Redondi, Pietro and Lini, Domenico (eds.), *La scienza, la città, la vita. Milano 1906: l'esposizione internazionale del Sempione*. Milano: Skira, 2006.

Romano Dall'Asta, Donatella, "La partecipazione di Brescia alla prima Esposizione nazionale di storia della scienza a Firenze nel 1929," in *L'Ateneo di Brescia e la storia della scienza*. Brescia: f.lli Geroldi, 1985–1988, 25–52.

Russo, Antonella, *Il fascismo in mostra*. Roma: Editori Riuniti, 1999.

Russo, Lucio and Santoni, Emanuela, *Ingegni minuti. Una storia della scienza in Italia*. Milano: Feltrinelli, 2010.

Rydell, Robert W. and Gwin, Nancy (eds.), *Fair Representation: World's Fairs and the Modern World*. Amsterdam: VU University Press, 1994.

Saggio bio-bibliografico degli scienziati della Lunigiana. La Spezia: tip. Moderna, 1929.

Salandin, Gian Antonio "Dal Teatro di filosofia sperimentale di Giovanni Poleni al Museo di storia della fisica," in *I musei, le collezioni scientifiche e le sezioni antiche delle biblioteche*, edited by Carlo Gregolin, 101–108. Padova: Edizioni dell'Università di Padova, 1996.

Salandin, Gian Antonio and Pancino, Maria, *Il teatro di filosofia sperimentale di Giovanni Poleni*. Trieste: Lint, 1987.

Salviati, Carla Ida, "'Sor Enrico'. Ritratto di un grande editore," in *Poggi e Bemporad editori per la scuola. Libri per leggere, scrivere e far di conto*. Edited by C.I. Salviati. Firenze: Giunti, 2007, 47–50.

Savelli, Dino, *La rotonda del Brunelleschi: storia e documenti*. Firenze: Esuvia, 1992.

Schroeder-Gudehus, Brigitte and Rassmussen, Anne, *Les fastes du progrès: les guide des expositions universelles, 1851–1992*. Paris: Flammarion, 1992.

Shapiro, Michael Steven (ed.), *The Museum: a Reference Guide*. New York: Greenwood Press, 1990.

Somsen, Geert J., "Science, fascism, and foreign policy: the exhibition Scienza Universale at the 1942 Rome world's fair," *Isis*, 108 (2017): 769–791.

Spackman, Barbara, *Fascist Virilities. Rhetoric, Ideology, and Social Fantasy in Italy* (Minneapolis-London: University of Minnesota Press, 1996).

Tarquini, Alessandra, *Storia della cultura fascista*. Bologna: il Mulino, 2011.

Taub, Liba, "Reengaging with instruments," *Isis*, 102, 2011, pp. 689–696.

Tega, Walter (ed.), *Guida ai musei di Palazzo Poggi: scienza e arte*. Bologna: Compositori, 2001.

Tomasi, Tina and Sistoli Paoli, Nella, *La Scuola Normale Superiore di Pisa dal 1813 al 1945*. Pisa: ETS, 1990.

Troilo, Simona, "National Museums in Italy: A Matter of Multifaceted Identity," in *Building National Museums in Europe 1750–2010*, edited by P. Aronsson and G. Elgenius, EuNaMus Report No. 1, 2011: 461–496.

Ucelli, Guido, *Origine realizzazioni e programmi del Museo Nazionale della Scienza e della Tecnica "Leonardo da Vinci"*. Milano: s.n., 1955.

Vinti, Carlo (ed.), *Alexandre Koyré: l'avventura intellettuale*. Napoli: Edizioni scientifiche italiane, 1994.

Name Index

www.ingramcontent.com/pod-product-compliance
Lightning Source LLC
Chambersburg PA
CBHW071322230926
51849CB00002B/293